DIRECTING VIDEO

By Thomas Kennedy

Knowledge Industry Publications, Inc.
White Plains, NY

The Video Bookshelf

Directing Video

Kennedy, Thomas, 1947-
   Directing Video

   (The Video Bookshelf)
   Bibliography: p.
   Includes Index.
   1. Video recordings—Production and direction.
I. Title. II. Series.
PN1992.94.K46 1989   791.43′0233   89-2738
ISBN 0-86729-172-9

10   9   8   7   6   5   4   3   2

Table of Contents

List of Figures

1 The Producer/Director's Role

The director of a small-scale video production plays many roles simultaneously, usually working as producer and director, and often operating the camera and coordinating the crew as well. This book proposes many organizational techniques that have proven helpful in making a production flow more smoothly. It is specifically designed for the director who operates in a "multi-tasking" fashion, performing many roles throughout the course of a production.

Part production manager, part creative designer, part group process facilitator, the modern director of a video project relies on skills generally not covered in standard television production curricula. These can be acquired from nonmedia-oriented sources including coursework in business, philosophy and psychology; practical experience dealing with customers (clients) in any extended business relationship; indeed, any past experience that includes managing teams of people.

In traditional large-scale production circles, typified by the Hollywood feature film, the role of the director has been seen as that of the creative force in the process. The same holds true today in the small-crew production environment, but with a major difference. *The director of a production staffed by a crew limited in size must act as creative force and must execute many of the roles that allow his creative vision to be fulfilled.*

He must plan every detail of the production day, oversee all technical planning, create and maintain a production schedule, and keep track of the individual takes throughout the production (whether during the shooting or by reviewing tape). He acts as much as producer as director, handling financial details and making decisions with major financial consequences, while managing technical logistics and making aesthetic decisions. Figures 1.1 and 1.2 illustrate the major difference between small-scale and large-scale productions.

1

Figure 1.1: A Small-Scale Production Crew

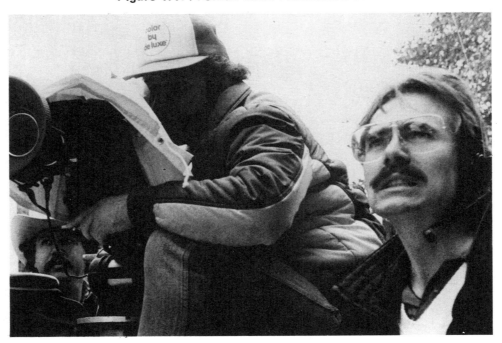

Figure 1.2: A Large-Scale Production Crew

THE TRADITIONAL APPROACH

In a traditional, larger-scale production, the director is supported by a sizable retinue of persons who fulfill a number of roles that allow the director to concentrate on the job of directing. An assistant director, or group of assistant directors, pre-plans the entire production sequence for a given production day based on their understanding of the director's needs and are responsible for creating a smooth staging of the day. They anticipate the next camera setup or series of setups and mobilize the crew members needed to create them.

Unit managers, production managers, location managers and production assistants tend to equipment and staging logistics, individual cast and crew needs, communications demands, transportation and meal requirements. The director's support staff takes care of managing the background extras, crowd control and location requirements.

On the technical side, cinematographers and directors of photography pre-design the production "look" with the director and direct their own "sub-crews" to create it. Gaffers, key grips, dolly grips, assistant grips, camera operators, camera assistants and focus pullers handle the details of lighting setups and camera movements.

Smaller-scale productions often require that sound be recorded "cleanly" on location or in the studio, since time and budget demands can preclude the opportunity to re-record dialogue. Sound recording requires careful attention to every possible intrusion by airplanes, trucks, dog barks, scratching materials around microphones, and background speakers. The multi-tasking producer/director is often asked to pass judgment on the acceptability of selected audio takes.

In traditionally structured productions this demand is also taken care of by a team of experts. A sound recordist selects the audio design, gathers the equipment and assembles a recording team that ensures that the right tracks are recorded in any given scene, at the right level. If any key elements are missed, they are noted and picked up for later "fixing."

Wardrobe changes, makeup application and maintenance, propping and styling are carried out by specialists or even teams of specialists in full-scale productions.

Continuity from shot to shot, logging of takes, slating and voluminous note-taking are handled by assistants. Clients present on the location or set are also "handled" by a retinue of persons with the appropriate people skills.

In sum, in the traditional, fully crewed production, the director, with all the traditional film-making resources at hand, is able to pre-plan a production and then delegate the execution of all tasks that might distract from the job of directing.

All this contributes to the director's ability to focus on even the smallest details of the directing task—noticing even the smallest mistiming and correcting it, changing an actor's inflection ever so slightly to make a better point, moving set elements in the frame or the camera itself, so that the best composition is created. In other words, refine the work of others and create the "director's touch" through those refinements.

THE SMALL CREW APPROACH

In the world of smaller-budget productions, many different forces operate on the director to make him less capable of focusing on details in the here and now. Economic realities and compressed production time-frames often force the director to perform many of the tasks that are performed by support personnel in the large-crew model.

The need to work in less controlled environments with a scaled-down crew requires the director to act as his own support group and perform many purely physical tasks. These may range from setting up the camera electronically to moving light stands and background props. These actions steal from time that could be spent sizing up the situation, rehearsing with actors to refine performance, looking at different angles to examine and change details that can add to the overall impact of a production.

The advent of videotape technology, with its immediate playback capabilities, has contributed to the tendency to do more with fewer people on a crew. For example, a director often chooses to act as his own camera operator as well as his own talent director, because he has the ability to check the quality of an actor's performance and the timing of his own camera movement by reviewing the recorded action on tape after each take. (On a film production, without "video assist," he would have to rely on his memory or the judgment of others.) Rehearsals on tape also allow the director to analyze the actions of extras and refine the directions given to actors and non-actors.

In much the same way, other crew members can double up on their responsibilities. Videotape operators act as sound recordists and video technicians. If mistakes are made due to this "multi-tasking," the errors can be caught and scenes re-done, while all resources are still available, after tape review.

Advantages and Disadvantages of the Multi-Task Role

Although immediate playback has been a boon to those wishing to create programs with limited resources, it has also created a situation in which the director can be overwhelmed by the demands of making many decisions at once. The result can often be a production in which the director doesn't step back quite far enough to see the whole picture. In effect, his directing role becomes less that of the "shaper" of the program and more that of the technician who calls "roll tape, action. . .and cut."

When a director has been acting as his own lighting director, gaffer, grip, assistant director, production manager, production assistant, "go-fer," etc., for some time, he may become quite good at balancing the many roles and at leaving no "loose ends." But often this is accomplished by using a formula approach. That is, in order to make no mistakes while performing several tasks at once, he restricts those tasks to habitual ones and does essentially the same show each time in order to reduce the risk of making a mistake.

Client Concerns

The temptation to aim low is particularly strong when the director not only is intimately involved in the production but also has to deal with a client during the production process. Many of the production activities that one is exposed to over the years make no sense to an outsider or novice who is not versed in the production process and its accompanying language. So the director with an inexperienced client present during the production often finds himself in the role of "explainer," which further detracts from his ability to concentrate on directing.

The need to explain the process becomes particularly acute when the director sees an opportunity to improve a project by making changes during shooting. The novice client is asked to "re-visualize" something he may have had a hard time visualizing in the first place. The director then must explain step by step what he has in mind, why it is better, and how the change will be made.

An even more delicate situation may arise when the director changes a production approach during shooting because he finds that the original concept simply won't work. For example, a series of cuts may be substituted for an extended dolly shot when an uneven floor makes the dolly shot impossible, or the focal length of the camera's lens and the shot framing may have to be changed to accommodate a nearsighted talent who has to work closer to the camera than originally planned.

Clients may feel that they have been betrayed, or that the director and crew aren't competent, or that the entire process has simply gotten out of control. At this point, the producer/director may spend an inordinate amount of time reassuring the client; the entire production suffers due to delays; and the client may feel that production values have been compromised, and that he is not getting his money's worth.

Making the Producer/Director Dichotomy Work

In these cases, the director can find himself in an extremely awkward situation because he is fulfilling the producing and directing roles. This is perhaps the most difficult of dual roles to manage simultaneously. But it is also one of the most common dualities to be found in a small-scale production. Whereas the director functions best when "absorbed in the moment," focusing on details and re-taking until the final product is as good as it can be, the producer functions best by "living in the future": making future setups go smoothly, staying on budget by not over-doing protection shots and all the while keeping an eye over his shoulder to sense how the client is perceiving the entire process, so that he might ensure that the client is pleased with the final product. These two functions seem to be in almost direct conflict with each other, and many people feel that they are, in fact, mutually exclusive.

However, in an in-house corporate audiovisual center or a small production company, these two roles may have to be combined and often are. You might ask, "Who in their right mind would ever want to assume these two roles at the same time?" Many persons in the production realm are motivated by challenges such as these and have developed some techniques that allow them to not only survive with their sanity intact, but even function creatively.

Build a Trust Relationship with the Client

One of the ingredients most essential to developing a smoothly functioning production environment that will allow a director to exercise creative options throughout the production process is creating a bond of trust between the client and the producer/director. The client must feel that he is listened to, that his option is valued, and that the producer/director has his best interests at heart. At the same time, the client should feel that the producer/director has the experience and judgment to make the best decisions, even if they might be in conflict with his initial intuition. In a word, the client should respect the professionalism of any director employed for a project.

Set the Tone

There are several ways in which this trust can be built. Most often, a client meets the producer/director early in the production process, away from any equipment and crew involvement that would directly establish that this person "knows what he is doing." There simply isn't the opportunity to let the client, or potential client, see firsthand how one commands the crew or frames shots.

Don't yield to the temptation to establish expertise by introducing technical jargon into these discussions, because this can create barriers that may prove disastrous in the long run to future communication. Clients may nod sagely at statements they only vaguely grasp and agree to production goals and methods they simply don't understand.

The director may suggest "a limbo setting with high key lighting" at the outset of a production meeting; the client may agree or defer to the director's judgment. But on the day of the shoot the director may find that the client considers this look too "artsy" for the executive or incompatible with the corporate image. A major regrouping may be needed to try to create a set with materials at hand that would be more to the client's liking. Needless to say, the resulting rush to create something (anything) close to what the client had in mind can result in bad feelings on all sides:

> DIRECTOR: Oh, you don't like the limbo look. . .Hey Jan, what kinds of flats have we got^

> PRODUCTION ASSISTANT: Two left over from that pseudo game show.

> DIRECTOR: What color are they^

> PRODUCTION ASSISTANT: Mauve and chartreuse, I think.

> DIRECTOR: That's fine. [to crew] Trot two softs in there and beam 'em up, kill the 1 K! It's gonna look great—no problem, no problem.

Be Straightforward

A better approach to initiating a trust relationship is to use straightforward language and a businesslike manner to define roles, avoiding the buzzwords of the production business, keeping an eye out for any possibly unclear language. It is easier to work together in a team approach if roles are defined as a part of the enterprise that you are both about to undertake, rather than as hard and fast identities.

It is often helpful to introduce yourself to the client at the outset in a manner such as:

> "Hello, I'm Zack Bogart, and I'll be working with you on this project. I'll be acting as director on the program."

The slight difference between stating "I *am* the director" vs. "I'll be working with you *as* the director" can have a major impact over the course of the working relationship.

Establish the Team Concept

You may wish to elaborate on the role of director in terms such as, "Once I get a clear understanding of the program you want produced, I'll be acting as your creative force. I'll be putting together the production team and working with you to get the most out of that team. I'll be selecting the production style and methods that will let us develop the best program for your needs."

This emphasis on working together as a team can help immensely as you create the budget and timetable that you need to control the production process. It can make it much easier to explain to the client your (mutual) need to establish firm dollar and calendar guidelines. In order to exercise your abilities as director, it is essential that you have a mutual understanding so that there are as few false steps in the process as possible. (In the smaller-scale production, nothing drains energy and resources as rapidly as having to go back to square one because of a misunderstanding along these lines.)

Set Mutual Goals

It is also helpful to begin defining the project in terms of what the client wants to achieve. As director, you do want to define goals and objectives in the traditional instructional design sense, but this doesn't have to take place in instructional designer language.

Listen to what the client imagines his or her program to be, and then followup with a simple, direct question such as, "When this show fades to black at the end and the TV set is turned off, what do you want your viewers to immediately feel (or know) that they didn't feel (or know) before the program began?" By listening closely for the first response—it may be one that is lost when the "objectives" and "long-term vs. short-term impacts" are discussed—you may discover the key reason for making the program in the first place. This may serve as an excellent guide as you design the production's style and pacing.

Granted, you are still asking what the objective of the program should be. But by not using the traditional buzzwords, you are telling the client that you are willing to listen to him in his own terms. You are also seeking a direct, intuitive response, which can be invaluable throughout the production process as it gives a clearer indication of the way in which the client thinks.

Uncover Hidden Concerns

Although many clients may be glib at the outset of a production, offering a rapid-fire description of who and what the program is for, it never hurts to probe for hidden objectives. For example, the client may give a grand description of all the aesthetic elements that need to be included in a program, without stating that the last project he did came in well over the targeted budget. The client's hidden agenda may be to see that this particular program needs to come in under budget (or he won't be doing any video projects in the near future as his departmental budget is up for review). You can suggest at the outset production approaches that meet the hidden objectives as well as the overt ones, making both you and your client "heroes."

Define Alternatives

The 18-location production schedule that a client initially describes may be a bit overambitious. You can serve that client well by suggesting methods of combining existing footage with newly produced segments to achieve the same general effect. Or, you may suggest a graphics treatment that utilizes existing artwork from collateral material without creating new, computerized (costly) graphics.

Clients often appreciate just such a common-sense approach from the director. They may, in fact, be more amenable to later suggestions, which might cost additional money but which the director feels are important to the overall production, once the director has demonstrated that he is willing to work *with* them to achieve their goals.

Find the Silent Client

Another approach, which serves the director well early in the process, is to probe for "silent clients." Recognizing that the person who commissions a program is not always acting totally on his own accord, you should ask him for a little background on the history of the idea. There may be a vice president who never enters the production discussions, but who is in fact keenly watching the outcome of the project; thus the director may find that he has more than one client. By knowing that the visible client has a need to please a "silent client," you will have an edge in developing a good trust relationship with the client with whom you must work directly.

Define the Viewing Environment

"Know your audience" has long been a guideline in the entertainment business. In the early stages of a production design, this cannot be emphasized too much. The director can create a strong bond with the client by showing a regard for the final use of a program.

Often, a client knows the audience in ways that the director cannot hope to learn in the short span of the production schedule. The client may be a sales manager who wants a motivational program for the sales force—a sales force that he has been a member of for many years. The director can reinforce the teamwork approach by the way in which he seeks the client's knowledge of the audience—asking the client's expert opinion of the types of programs (video or other media such as meeting formats) that have worked well or poorly in the past with the same sales force.

Positive Client Positioning

"Position" the client by further defining what his role is in relation to the director. Let him know that his input is valued, but that the director needs some room to operate in his own sphere. It can help to offer to sit in on pertinent meetings and to review materials that the client hasn't brought or thought important but that you feel may help you to gain a greater understanding of the people for whom the program is intended.

Clients are often pleasantly surprised when a director seeks to know as much as possible about the "when" and "how" of the viewing situation. You might ask, "Will this program be seen on employees' own time, or will they be required to view it at a specific point in their workday?" Explain that it will help to give you a better understanding of their frame of mind and so will guide you in selecting the timing or placement of particular program elements. It may even assist you in determining the most effective program length. Showing such a regard for the overall impact of the production can serve to strengthen your credentials at the same time that it emphasizes that you have the client's concerns at heart.

Create a Favorable Initial Impression

The overall impact on the client of initial meetings should be that the director is not the practitioner of some arcane art, but a professional who attempts to understand the client's need as well as possible, is willing to learn from the client and will be working with the client to create a program that is *effective* from the client's point of view. If this impression is made and maintained, there will often be a remarkable absence of conflict throughout a production. This can have the net effect of freeing the director to *direct* the program with the support of the client, rather than struggling with the client for control.

It is much easier to work with a small crew if the director has this sort of client support. Without it, too often he can feel torn between the moment-to-moment demands on his attention during the actual production and the need to constantly "look over his shoulder" to check on the client's current state of mind. This can become particularly acute when the specifics of a shoot require quick thinking on one's feet.

Anticipate the Unexpected

Since productions don't always proceed perfectly according to plan, the ability to think while doing is often essential to their success. A positive director-client relationship also serves very well when the director sees a creative opportunity and must ask the client to trust him. The ability to elicit such trust quickly can often add greatly to the overall production value of a program. By building a strong client relationship, the director can be freed to follow his intuitions and take advantage of creative opportunities as they arise—to act more fully as a director.

2 Understanding the Production Need

The initial meeting with a client is a director's first opportunity to begin to understand the production need of that client's program. Here, the director begins playing the dual role of producer/director. He must begin to put together the basic structure of a production approach, budget and timetable. A successful first meeting is a key step in building a positive relationship with the client. But this is typically a meeting in which several forces are at work.

The client may have reservations about the advisability of doing a video program in the first place, or he may be hesitant or unable to give anyone the "whole picture" of a project at this point, since it may still be vague in his own mind. (This will be particularly true if the client isn't familiar or comfortable with the production process.) This may be the client's first opportunity to meet the producer/director and he will base much on his first impressions.

Getting accurate, complete information in the course of this meeting can pay tremendous dividends. It allows you to begin the organizational and creative thinking processes that are crucial to making a successful, effective program. The next contact with the client, whether by telephone or in person, will be greatly enhanced if your basic questions are answered and you can offer thoughtful, creative suggestions based on those answers.

Bringing a basic agenda to such a meeting will greatly facilitate this process of information gathering. It will also help to reassure the client that he has chosen a person who is competent, thorough and resourceful, as well as creative. This helps in establishing the trust relationship so crucial to a small-scale production.

Figure 2.1 is a suggested checklist for such a meeting. But first we will deal with the specifics of this type of meeting, focusing on some of the most common questions that arise and some potential "sticking points" that must often be dealt with.

Figure 2.1: Initial Client Meeting Checklist

1. Elicit objectives: probe for hidden goals

2. Identify (and establish review process for):
    Policymakers
    Content experts
    Legal/proprietary concerns

3. Define:
    Audience
    Viewing situation
    Support materials
    Distribution needs

4. Establish a production timetable:
    Script approvals
    Casting procedures
    Artwork and graphics development
    Production dates

THE BUDGET QUESTION

Often a novice client's first question to a producer/director will be, "How much will this program cost?" This is understandable. Clients may not be familiar with production processes and terminology, but they can readily understand the bottom-line dollars and how these relate to the overall budget they may have allocated to a project.

Don't fall into the trap of answering such a question quickly! Yes, the client expects you as the expert to have the answers to all his questions; you may feel that to answer the question indirectly could erode your credibility; other persons at the meeting may be looking at you expectantly. However, a quick response to such a question can create major problems later on in the production process. Without knowing many of the client's specific expectations, a quick answer may not include many production elements that the client may have in mind. This can only lead to disappointment or dispute at a later point.

Since there is rarely enough information at hand at this point to answer such a question well, I have found it helpful to have a selection of standard approaches available. The key is to be sincerely helpful and not to appear uncomfortable in the face of the question, which is a valid one.

One approach is to note that "I have done several programs of this nature and they have ranged in cost from X dollars to Y dollars." This immediately raises the question in the client's mind: "Well, what made for the difference in cost?" Before the question is asked, though, my further response is: "Differences in production needs and production values make the major difference in production costs."

This technique does not beg the question. Rather, it redirects the discussion to an area in which you are able to ask for more information from the client, without putting him on the spot. This allows you to probe for as much information as possible about the program objectives, the general viewing context, the typical viewer profile and his immediate viewing environment.

Budget versus Program Objectives

I explain that only with a clear understanding of the production need will I be able to estimate the best use of the client's dollars. The client's overt objective may be to inform employees about health care and employee benefits. If that is the program's primary and sole objective, it can be produced in a straightforward manner for a basic budget.

But if that same program has as a secondary objective to make employees feel that their employer is a *leader* in its field and provides *quality* care for its employees, then the program should contain high production values. Often, it helps to explain in this process that the viewer judges the validity of a stated message by whether it is "wrapped" in a production style that supports that message.

"Tailoring" the Production Style and Budget

By contrast, I may use the example of a president's message that is intended to explain the need to reduce production and lay off a portion of the work force. Such a production should be well designed and the message should be delivered clearly, but an ornate production design would undercut the sincerity of the message.

The purpose of this is to gently educate the client about the value of completely understanding the production need before focusing on a specific dollar figure. There is, of course, an element of self-protection involved in this approach, since a hasty answer to the question of cost will stick in the client's mind, no matter how much qualification follows that dollar figure.

It is better to establish a qualifying context at the outset so that you are not haunted later by a dollar figure quoted in haste. Also, if handled well, this exchange can serve to establish the fact that you, as the producer/director, are flexible and can work within various budget ranges—that your primary concern is in understanding and meeting the production need, rather than in creating a program with as much "glitter" as possible.

The Viewing Context

In order to understand the overall scope of the program, you need to define the general viewing context. Does this program stand alone, or is it part of an ongoing flow of information? Are there written materials that precede, accompany or follow the viewing of the program? If so, what should be the role of the materials in relation to the program? Are there presenters or facilitators who will lay a specific context for the program?

Only if these questions are addressed can you, as the producer/director, begin to understand the production need in terms of its overall scope. Only then can you begin to see what content needs to be included in the program versus what can be provided with print or an in-person facilitator.

Technical Considerations

If a program must be viewed as it is projected onto a large screen, the technical aspects of the video and audio recording must be of a higher quality than if the program is to be viewed from a VHS tape on a thirteen-inch monitor, or on a portable viewing system with a five-inch screen. These considerations will dictate whether the production should be made on one-inch or on three-quarter–inch videotape, and what quality of camera must be used.

If the program's sound track will be heard on large speakers in a large room, or on private headphones, the track must be handled more carefully throughout the production process. These may seem to be purely technical concerns, but they translate into additional dollars being required for the production.

Again, if these questions are raised at the beginning of discussions, the client will have a much better feeling about the process than if you were to quote a figure based on a typical program of this type, only to revise it upwards when these concerns are raised later.

The Sample Reel

One technique that has proven invaluable over the years has been to build up a sample reel of similar types of productions, with varying production values. I have shown clients an excerpt of a tape that utilized existing slides as graphic support, and followed that sample with an excerpt that used digital graphics created specifically for the program. The client immediately sees the difference in readability and aesthetic quality: the graphics "made for TV" may be four or five times as expensive as the cost of transferring slides to tape, but the difference translates into more effective communication immediately evident to the viewer.

Clients can readily appreciate the difference between non-actors and actors when the performances are contrasted side-by-side in excerpts from finished programs. They can see the difference between cutting from one shot to another versus well-executed dolly shots when you show real examples and point out the techniques. They can hear what happens when you use library music that sometimes fits the action well and sometimes just "lies there" versus using music that was post-scored. The post-scoring *underscores* and enhances dramatic action, builds a sense of action and wonder, and allows visuals to have more power than they would have otherwise.

The Value of Samples

Real examples of programs with marked differences in production values allow the client to participate in informed decisions about the best use of his production dollar. In the course of the project, it allows the producer/director to recommend one production style over another for aesthetic reasons, without creating an adversary relationship in which the client questions every expenditure. It allows the client to know what he is giving up or trading off by choosing the less expensive of two options and, by the same token, see what he is gaining when he agrees to a more expensive production style.

The use of such samples serves to lay a healthy groundwork for discussing production costs. It removes much of the mystery that too often surrounds the producer/director's role and reinforces the trust relationship with the client.

Ongoing Review of Client Expectations

A second effect of using real examples is that the client begins to review his initial expectations. As he sees questions regarding costs develop into discussions about style and aesthetic value, he begins to refine his original (and possibly vague) idea of a program. His imagination may be sparked by real examples and he may offer more specific suggestions about the production, which give the producer-director a much better idea of what his client wants.

This lets the producer/director begin to work with his client in terms of what the program will look like rather than simply what it will cost and what it should achieve, and it opens the door to creating realistic expectations.

THE PRODUCTION TIMETABLE

A second client expectation that must often be dealt with at the outset of the pre-production process is the production timetable. Since we are all exposed to television programs in our homes that are produced quickly and smoothly by armies of invisible production personnel, we can be unaware of the vast amount of planning and preparation that is involved in their creation.

The producer/director should set as a major task of any initial client meeting the development of a *basic* production schedule. This does not have to involve detailed steps for every production element but should include script development stages, production planning and review steps, tentative production dates and last, but not least, the required delivery date for the finished program. (Figure 2.2, at the end of this chapter, includes a sample production schedule.)

True Delivery Dates

Since the delivery date is the determining factor for many of the other steps, it makes sense to probe for the true "drop-dead" date for a program. Clients who are used to vendors who are chronically late in delivering products may tend to give false deadlines. This may mean that they say they need a program two weeks sooner than they really need it, to allow themselves a cushion in case you are late delivering the finished program.

On the other hand, since the client may not be familiar with all the steps involved, he may say only that he needs the program on a given date, without explaining that, in fact, he needs 300 copies in PAL and SECAM format to be delivered to various points overseas on that date! This is also a false deadline for the producer/director, if taken at face value.

You should ask very pointed questions regarding: when a program needs to be available for review, whether the persons involved in the review are guaranteed to be available, whether a legal department needs to have final say, what the duplication requirements are, and whether any shipping requirements are involved.

As director, you want to use every moment effectively in the course of the production. If you find out that your deadline is earlier than expected, you may have to hurry up the later steps in the production more than necessary. Often the editing benefits from an occasional "step back" that allows you to regain a sense of the whole; this is the first thing to go when the production timetable becomes accelerated on short notice.

Continuing Client Involvement: Program Reviews

On the other hand, a false early deadline can create an extended review process in which too many people become "Monday morning quarterbacks." Those persons not involved in the production of a yet-unreleased program may create an overwhelming pressure to rethink and change the finished program, not always for the best of reasons.

They may simply feel that they should have been included in the program development process, or that they have a particular insight that will make a major difference in the finished program. A common statement may sound like: "Since the program isn't needed for another week or two, it should be easy to change just that one line." Their feelings may be valid, or they might simply stem from a feeling of having been left out.

Upon hearing such comments, the immediate client may feel unsure about a program that hasn't been publicly released and may overreact to the pressure even when the recommendations run counter to the intent or style of the program. This puts the producer/director in an awkward situation. Your client may ask you to judge whether he should remain loyal to your original production design or respond to his co-workers' concerns.

Developing Appropriate Input

The best production timetable is one that allows for comments and input from the appropriate parties at appropriate points in the production. To this end, it is essential to identify those "appropriate parties" and their roles early in the project—ideally by the end of the initial meeting with the client.

One effective approach is to ask the client to identify and categorize those who should be included as one of three types:

- Policymakers—executives who should review script for consistency with company image and policy. Often, these persons may need to be included for courtesy reasons.
- Content Experts—those with technical expertise who can identify technical or factual errors and correct them. (In technical training programs, for example, the content expert may have to review the script, actually be present as scenes are shot that demonstrate technique, or simply review an initial edit.)
- Legal Resources—those whose job it is to review company publications for adherence to copyright and trademark guidelines or sensitivity about proprietary matters.

The purpose of including these persons early in the process is obvious. If they have the power to require changes, it is much better for the project that they suggest changes when they can be accomplished on a word processor rather than when such changes require calling back the talent to re-record the script.

Developing Role Definitions

The purpose of the specific categories is less obvious, but quite useful. By defining the roles of each reviewer, you limit their intrusion into areas in which they can contribute to the consistency or accuracy of the program. Too often, an executive who is simply asked to "review this script" will feel compelled to rewrite it in his own words, the way he would deliver a speech or letter. But if he is asked to review a script "for consistency with company policy and positioning," he will tend to focus on that specific area.

Content experts who are not given a defined role may feel that they should contribute more than is your intention. In the worst case, they may begin to vie for control of actors and performance—unintentionally attempting to usurp the director's role.

Legal experts may take perfectly clear spoken language and turn it into indecipherable "legalese." Their training can lead them to make every sentence "perfectly clear," covering all cases past, present and future imaginable. Often the result is a script that cannot be translated well to the screen because the spoken word is quite different from the written word.

By defining these roles (or asking your client to do so) you allow reviewers to be included in the process and contribute their expertise while heading off potential conflicts. In addition, you create a greater level of interest and enthusiasm for the project in other areas of the company than you might otherwise.

Maintaining the Production Schedule

The inclusion of appropriate parties in the production timetable also allows you to enlist the client's aid in keeping the production on track. By probing for the needed involvement of these persons and stating the need to gain their approval at the right time, you set the stage for a give-and-take. You can draw up a schedule of approval stages or "sign-offs" that allows you to proceed on schedule.

In other words, you promise to deliver script, casting tapes, set designs, etc., on given dates. At the same time, you require the client to provide appropriate approvals by given dates. You can state this in a way that the client realizes that he is, in effect, entering into a contract with you in order to produce the program by the due date. If the client misses his deadlines for approval processes, then you should not be expected to meet your production deadlines on schedule. You can explain that a given amount of time is required on your part for each step and that, if an approval date is missed, the production will either be set back by his missed deadline or the overall production quality will suffer since you are not being given the required amount of time to perform your tasks in the best manner.

Making the Hidden Visible

Sometimes it is necessary to explain the amount of "hidden" work that goes into a production. You may need to explain to the client that you cannot cast actors for a program until you have an approved script. And further explain that the casting process involves several steps. You will review talent head shots or composites and pre-select actors for visual approval. After approval of actors for the right look for the company's self-image, you will then proceed to auditions, from which you will select audition tapes for approval. After that approval, you may do callbacks and will have to check actors' availability for the production dates.

Once you have explained any one task in terms of its requirements on your time, most clients will begin to see that you are going to be spending quite a bit more time on their behalf than they had originally realized. More often than not, this will motivate them to do their part to meet their deadlines.

One device that I have found helpful is to record the finished script in my own voice (a scratch track) on an audiocassette that allows me to listen to the timing, phrasing and pacing of the script while I am driving or doing other tasks. I will explain to the client that I like to have the final approved script at the earliest opportunity, since this allows me more creative time to begin to visualize various production opportunities, consider who might be the best narrator, what action might be most appropriate and what music might best enhance the script. In effect, it allows me to work harder for the client to provide him with the most creative program possible.

Missed Deadlines—and Consequences

The other side of the picture is to point out that, if a given approval date is missed, we might still be able to meet a given deadline, but it will involve contingency planning. This may mean shooting or editing on a weekend or at night, with increased costs. Quite often these working conditions lead to less than optimum creative choices as well. The implication is that if the client wishes to stay on budget, then he must also stay on schedule.

So much emphasis is being placed here on timeliness because the smaller-scale production usually has less margin for error than the large-scale project. Although a feature film may go over budget because of delays in production and become a better picture that makes back more money at the box office, a program that has a fixed, non-paying audience cannot hope to recoup the cost of delays through popularity. A newsmagazine-format program or new-product marketing program may, in fact, become perfectly useless if it is delayed to the point where it becomes old news.

Probing for Additional Information

Discussions of budget and timetables provide the perfect opportunity to discover whether the client has other concerns about the project. This lets you know immediately what production styles or techniques the client might find inappropriate.

Ideally, you should leave the initial meeting with a clear understanding of the agreed-upon program objectives, a definition of the audience and viewing situation, a good sense of the intended general tone of the program, as well as the optimum length and required delivery date. This will allow you to follow up the meeting effectively.

THE DIRECTOR'S PERSPECTIVE

All the above concerns regarding timeliness, keeping to a production schedule, meeting deadlines and gaining sign-offs may appear to be a form of pessimistic planning. However, the director who embarks on a project with a firm deadline for delivery and a fixed budget must realize that time is always his enemy. Creative decisions are often made in a playful mode, when one's primary concern is not the pressure of time and when there is the latitude to try different alternatives.

By buying time at the outset of a production, the director creates a greater possibility that he will be able to exercise his own creativity. Only with the time to "play" can a director use his full talents.

Follow-up: The Proposal

Before proceeding with a thorough production design, the producer/director should touch base with the client to ensure that he has a clear understanding of the production need. This should be done in written form so that both parties can easily address any misconceptions. A simple letter may suffice, although in many cases, a more formal proposal or "platform statement" may be appropriate.

A well-written proposal can also assist the client in explaining the project to other persons within his organization. If his initial budget estimate was too low, and he would like to adjust his budget rather than his program idea, this document may also assist him in securing additional budget dollars. The formality of the proposal will also serve to assure the reader that the production will be executed in a professional manner.

Proposal Design

A proposal should identify who is commissioning the program, what its intent is (primary and secondary objectives), and give the reader a sense of the finished program. This may be a rough treatment or a general description supplemented by a sample of a possible scene. A budget range and timetable should be included as well. Figure 2.2 provides a sample proposal.

Figure 2.2: Videotape Production Proposal

PROJECT: Impact '87
CLIENT: Stylistic
CONTACT: Linda Murphy
TELEPHONE: (415) 123-4567
DATE(S): April 12–15
LOCATION(S): Phoenix Convention Center
EDIT REQUIREMENTS: Betacam SP to 1 inch

THE PROJECT

We will create a 4:00 to 5:00 videotape which can be utilized by Stylistic franchisees to introduce thier employees to the new fashion cuts that Stylistic will be promoting in 1987. The tape will be done in a fashion runway setting to emphasize the new fashion orientation of Stylistic's current marketing strategy, to visually stimulate viewers and to promote their perception of the Stylistic organization as an exciting place to work.

THE PRODUCTION APPROACH

We will videotape the Impact '87 presentation at the Phoenix Convention Center on the two successive nights that it is presented, in an identical format each night. Utilizing three isolated cameras each night will allow us the variety of angles that a six camera shoot would present.

The program's soundtrack will be an original composition commissioned for Stylistic— *Superstyle*—sung by Mark Robson, the composer, with his own band as backup. We will record this on location for the proper sound presence and "live feeling."

The finished program will include quick cutting between Mark onstage, models in full runway shots, closeups of hairstyles as worn by each individual model and medium shots as models make their runway turns. The overall effect will be of a lively fashion show with many strobes in the audience evident on camera.

PRODUCTION SCHEDULE

March 30	Pre-production meeting—review staging and camera placement, begin crew booking
April 2	Location scout, review requirements with hotel
April 6	Record original music scratch track for timing
April 7–8	Time music, design shots, design lighting
April 9–10	Prep equipment, pre-production with crew
April 11–12	Travel and load-in to hotel

Figure 2.2: Videotape Production Proposal (Cont.)

April 13	Pre-light, place cameras, rehearse
April 14–15	Tape production
April 16	Strike and travel, submit original tapes for window dubs
April 17–20	Edit rough-cut
April 19	Review rough-cut with client, revise, if necessary
April 20	Edit on-line, make program dubs

By providing such a document, the producer-director proves to the client that he is a good listener and can assimilate information well. He also gives the client a sense of his ability to translate the client's ideas into the first phase of the production process. (Bear in mind that this is not a script, but rather an overview of the production. It should leave any reader, but particularly the client, with a feeling for the style of the finished production.)

Since many programs have similar elements, this task can be accomplished easily by working from a "boiler plate" format. Elements of each proposal can be incorporated into succeeding similar proposals, and general statements tailored to the specific need. Obviously, word processing programs are particularly helpful in this process.

The net effect of a well-prepared proposal (that is accepted by a client) is to further cement the client-director relationship. It acts as an informal contract that allows each party the opportunity to agree on an overall approach for a project. Any discrepancies between client and director's perceptions are likely to be highlighted by this process. It affords a clear understanding for both parties and allows them to embark on the production process knowing that they are each thinking of the same general program.

3 Developing the Production Design

In any production, the director is responsible for ensuring that what is put on tape can be edited into the program called for in the script. This is not the same responsibility as that of a writer, but follows the writer's completed effort and sometimes involves some interaction among the producer, writer and director. The writer's area of expertise lies in putting on paper a concept with its corresponding scene descriptions and dialogue passages. Often, the writer is asked to "write to the budget" which means that he must only call for scenes that can be staged within a given amount of dollars.

A script that is written to a budget will be restricted to a given number of production days, with a limited number of actors and scenes. The director may begin to visualize the finished program and select the resources that will allow him to deliver the needed footage within the budget parameters. (There is some overlap and interaction between producer and director roles in this regard, but for simplicity's sake we will treat the director's task as if it includes any necessary discussions and feedback with the producer, particularly since the producer and director are often the same person in smaller budget projects.)

SCRIPT DEVELOPMENT

A director is sometimes included in the script development process, usually at the client's discretion.

When making a commercial, for instance, the director may have little input in the script development process. The script for a commercial often reaches the director after a long development process with many approval cycles, and the director is only asked to execute an approved, detailed storyboard. For a director to suggest substantial changes once a commercial script or storyboard reaches his hands is often inappropriate, since such changes may require a complete recycling of the script or storyboard through the approval process, often with unpredictable results.

25

In longer-format programs, such as training or marketing videos, the director has greater latitude for such modifications. However, it is understood that the director must adhere to the basic intent of the script and the budget parameters. Since these programs are geared to a fixed-size audience, creative changes that cost additional budget dollars cannot be justified solely on the basis that they will make the program more popular.

Because these programs have a very defined audience and limited viewing life, their effectiveness must be measured differently than that of a general-release film. In short, creative ideas are not justified solely by the fact that they will produce greater viewer interest or viewing demand. The final standard for judging such programs will always depend on the specific objectives of the program. For example, a marketing video which has as its purpose the introduction of a telephone system will be judged on the basis of several factors other than a final increase in system sales. It will be judged on whether it clearly presents a product's benefits first to the sales force and ultimately to a customer-viewer. It will be judged on its acceptance by the sales force. Does the sales force feel that it is a strong sales support tool? Only indirectly will it be judged on its effectiveness in increasing sales as this can only be measured indirectly.

However, the director is typically given great latitude simply because of the length of the script and the timetable for producing the program. To completely storyboard a longer-format program shot by shot is a larger task than the typical budget permits. Written descriptions of the action and storyboards of key frames within scenes are more common than comprehensive shot-by-shot storyboards.

The organized director does, in effect, create a rough, comprehensive storyboard of the program in his own mind prior to developing his shot list, but this is often kept fluid until he has determined all his production parameters, and often involves several alternate angles for given shots which will be determined by other factors including performance and editing choices.

Script Formats

In longer-format programs, the director may receive a script from the writer in manuscript form, or in split-script form. There are advantages to each.

The Manuscript Format Script

The manuscript format provides a general description of each scene, followed by dialogue and a narrative-style description of action within the scene. The writer who uses this style leaves the specific breakdown of shots within the scene to the reader's (i.e., director's) imagination.

The manuscript format allows a quick reading of dialogue and narration intermixed with action and prop considerations. This allows the director to freely associate his own vision of the appropriate points at which camera angles will change. Very little is said about those angles unless the writer has chosen to punctuate specific lines of dialogue or narration with a shot that he feels is critical to communicating his own vision of the scene.

For example, Figure 3.1 shows a portion of a script in manuscript format in which action and dialogue take place to set the scene for a Western-style confrontation in a barroom.

Figure 3.1: A Manuscript Format Script

INTERIOR: SALOON—DAY

Four patrons are at the bar talking quietly, along with the **BARTENDER**. They are a **BANKER, NEWSPAPER EDITOR, SHOPKEEPER** and **RANCHER**. They have drinks in front of them.

RANCHER: You know what really gets me. . .

BARTENDER: Shhh. Keep your voice down.

With a nod he indicates someone (OFFSCREEN) whom he doesn't want to overhear the conversation.

RANCHER: I moved to this territory 'cuz I figured I'd have more choices
(quieter)   out here, you know, more options. But it's the same here as
                everywhere else.

NEWSPAPER   Tell me about it. I hear the same thing over and over.
EDITOR:

BARTENDER: What about you, Frank? You've done business with some of them
                other fellas.

In this format, it is very easy for a director to pick out what characters are required because they are capitalized—and key props and actions are noted in a manner that lets him see what is needed in the scene to carry the writer's intent—but the flow of the dialogue is uninterrupted by notes regarding angles, camera movement, or detailed action by the characters. The reader who is experienced in production methods can create a list of production needs. He is not presented with how the scene should be shot angle by angle.

This format has the advantage of allowing each reader to visualize a scene in his own fashion—a boon for the director who wishes to add his own ideas regarding visual emphasis and counterpoint. However, it can have some disadvantages for client presentations. The inexperienced client may focus strictly on the dialogue and miss the nuances and details that will accompany the finished production. For such clients, a split-script presentation can be helpful to convey a better impression of that finished program.

The Split-format Script

A split-format script is practically essential to thorough pre-planning of a production when full storyboards will not be created. This format requires that the author (whether writer, producer, director or a combination of these) notes at what point a camera angle changes and what is encompassed by each angle. The same script that was presented in Figure 3.1 is translated into a split script in Figure 3.2.

Figure 3.2: A Split-format Script

Wide shot from saloon doors to interior of saloon. Includes all 5 characters	**RANCHER:** You know what really gets me?
Two shot: Rancher and Bartender.	**BARTENDER:** Shhh. Keep your voice down.
(dolly to include Newspaper Editor)	**RANCHER:** I moved to this territory 'cuz I figured I'd have more choices out here, you know, more options. But it's the same here as everywhere else.
	NEWSPAPER EDITOR: Tell me about it. I hear the same thing over and over.
Close-up Bartender	**BARTENDER:** What about you, Frank? You've done business with some of them other fellas.

In this first breakdown of the sequence into discrete shots, the director can begin to choose the specific visual elements that he feels will enhance and underscore the writer's dramatic efforts. In succeeding script revisions, he can use the left side of the script to keep track of his additions to the visual staging and the right side to keep track of his additions to the audio track.

Throughout this process the director should be guided by the basic structure that he finds in a thorough reading of the script as a whole. In any script there are traditional elements that make that script successful. The director's job is to identify these key elements and to bring them to the fore through his casting and staging choices.

KEY SCRIPT ELEMENTS

In a dramatic script, the director must identify the main character or characters and understand why these characters act as they do. He must find, through his script reading, their motivation. This springs from two areas.

Prior Characterization

First, the characters had a life prior to the action within the script. The director should find ways to stage action to underscore the result of this prior life, so that the viewer gains a quick sense of that life. Often, the writer will provide some of this information in the form of a written description (a must in a traditional screenplay). A character can be described in terms of age, occupation, family background and key events in his childhood and recent life history.

For example, in a script I directed for United Way, the main character is described as a single, energetic male, in his early thirties, a computer programmer who is dedicated to his work and concerned about others. Since the program was a short-form campaign film running less than twelve minutes, I had to find ways to give the viewer a quick "take" regarding that character. The program opening was designed as a montage in which we see this character leave his home for work on a typical workday. The writer's purpose was to give the viewer a sense of the character's occupation and the geographic area in which he worked.

I chose to add elements supporting the character's background. He didn't simply get dressed for work, he dressed while he finished printing a file from his home terminal and simultaneously blended a health drink. He didn't simply hop into his car, he greeted the newspaper delivery girl and gave her encouragement on his way to his car. These small additions were in character—and more than that, they served to flesh out the character in a short amount of time without requiring additional dialogue.

There are many ways for the director to accomplish this, but it requires a thorough understanding of the character based on a complete script reading; additional research (e.g., what would a typical programmer who was caught up in his work do?); and the director's imagination (how can I show someone to be "a nice guy" quickly?).

Current Characterization

During the action in a program, characters are called on to interact with other characters and their environment. The director should choose staging elements that will flow naturally from the prior character that has been established, and will also flow naturally from the actions that the character takes within the script's development.

If a co-worker greets our main character in a distracted fashion, and he is the concerned type mentioned above, he should take note of this through his reaction. If he were established as a more single-minded, driven type, he would ignore the co-worker's state of mind and simply acknowledge the greeting.

There may be no difference in the actual written dialogue exchange between the two characters, but it is the director's task to understand his key characters well enough to be able to answer the question: "How would he (she, they) react to this event?" What reaction or expression would accompany this line? Would this character be doing something else throughout this interchange or would he stop what he's doing and devote full attention to his response? By choosing appropriately, the director ensures that the characters will have consistency throughout the program and that the characters will ring true.

Primary Actions

In any script, whether instructional, informational or dramatic, there are always a few key informational points that guide the overall production. The director should identify these and choose his staging options appropriately.

For example, in an instructional video about operating a new telephone system, the director should identify what is new and different about the system, or what key points should be learned at this phase. These points should be distinguished from basic background information or contextual information, and the director should choose his staging to highlight these elements.

If the key point about a new telephone system is that it lets an operator do several tasks at once, the director can choose action and angles that emphasize this point clearly to the viewer. He may choose to open the program with an operator fielding several calls simultaneously, intercutting console close–ups of the lights that indicate each call with shots of the operator working in an efficient, relaxed fashion. This is a simple example, but identifying key points will allow the director to choose appropriate shots easily and make the client's and the writer's intentions come fully to the screen.

To accomplish this, the director should take any informational script and rank its informational points in a fashion that will allow him to establish a tiered system, then, from that ranking, isolate three or four specific points that will rank as primary. In our phone example, the system may allow voice and data recording, hands-off use with voice-activation, integrate easily with existing systems, free operators to perform several tasks simultaneously, fit easily on a desk top, and so on.

Often scripts developed in-house in a corporate environment or edited heavily by in-house managers will include many subagendas important only within the in-house client's own environment. Key points and objectives are often obscured by the time the director receives an approved script. Other scripts are developed by writers whose primary experience is in print media and who do not write clearly for the screen. As a result, the director is often called upon to help re-clarify the original intent of a video, pare down a script to a manageable size, review it and "massage" it for the best translation to the video screen.

The director should make educated guesses as to the primary ranking of operational or design features or benefits and present them to the client. If necessary, he should enlist the aid of content experts, defined earlier in the process, and get a consensus on these key points to develop an approach which highlights the key points. Only then can he orient the production design to underscore what is most important. Without this sort of breakdown, he risks a program without structure and one that may generate little viewer interest.

Maintaining Viewer Interest

In the barroom scene described in Figures 3.1 and 3.2, the director is choosing to hold back the identities of the patrons at the bar by shooting them from the back as a group in the wide shot to set the scene, then revealing the individual speakers, one by one, as the scene progresses. This allows him to control the movement of the scene on a basic level through the gradual release of information. This creates a continuing interest in the scene for the audience, a desire to see more. This progressive unfolding of the characters' identities is one way of carrying the scene forward. In the first steps of script breakdown, the director will look for devices like this one which are consistent with the script content and maintain continuing viewer interest.

Another device is the movement of the viewer through the scene. In the example above, the director is choosing to add that movement in two ways. One, the chosen camera angles, when cut together as edited shots, will move the viewer through the physical space of the barroom as the action unfolds, and the actual camera movement within a given shot will move the viewer's perspective from one point to another. Two, the director could choose to add character movement within the frame, either of principal actors or of extras, so that there is always new visual information for the viewer.

THE DIRECTOR'S ENHANCEMENTS

These are the special additions that a director is hired for *beyond* his ability to direct the talent performance itself. The director will attempt to enhance the writer's work with a variety of methods, while remaining consistent with the basic thrust of the script. Often, he will suggest additional ideas as he continues to prepare the production.

For example, while auditioning and casting, the director may discover new ideas of character interpretation and possible bits of business that characters might do. Sometimes, the audition process will also reveal the limitations of the talent available. (For example, in the program described above, a classic white-hat hero is called for to play the part of U.S. Sprint. Such a part is often difficult to cast well in many areas of the country.)

In this instance the best actor in terms of visual type had the least dramatic experience, and the director chose to scale back the number of lines required for the character. The villain, AT&T, was easier to cast and I decided to allow this charater to carry more of the spoken lines. The net result, a Gary Cooper-like hero who was strong in his silence, played quite well within the original intent of the script.

Location Considerations

When a script provides general descriptions of locations, the director should give a great deal of thought to the elements of a location that he feels will enhance a production. He should read through the script and develop a good sense of what are essential and what are optional elements for a location. Naturally, a larger budget with a long pre-production allows him to adhere more closely to his choices, whereas a short pre-production or limited budget requires him to be more flexible.

As the director visits locations, he becomes aware of available visual details, such as the specific design of a set of saloon doors, that can become foreground objects to lend more interest to the frame. He reviews available props and set objects to determine what will enhance the production with authentic details and can begin to see the finished "look" of the production in his mind. As he does this, he can embellish the script with those details and, in many cases, flesh out the script so that the client can begin to share his vision. The writer's script becomes far more than words on the page and begins to take shape as a viable script.

Early Idea Exchange

If the director has the budget support to allow early interaction with set designers, prop people or lighting and sound persons, he can begin to suggest ideas as they occur to him and enlist their aid in gathering appropriate props, developing the right sound elements, pre-designing a visual look through filter and lens choices, etc. However, if such interaction is not possible due to a small pre-production budget, the director may wish to use the split-script format to begin to accommodate his own ideas as they occur to him. Incorporating these ideas in revised split scripts ensures that they won't be overlooked as the production progresses.

For example, in the above script, I saw an opportunity to include an additional character, a helper who sweeps up around the bar. The early part of the script involved long exposition through dialogue, during which the main characters would have little physical action. By introducing the bar helper, I could add motion—he could cross the back of the frame as he swept up. This addition also created opportunities for the characters to react visually to what was said so that the back and forth patterns of the dialogue could be broken. The development of my production design could then include this character as a foil who would provide richer opportunities for editing scenes.

Pre-Staging: A Mental Exercise

In the next step, as I saw the actual barroom that was to be used in the scene (a portion of a theme park during the park's off season), I could begin to place my characters in the location, then visualize the opportunities for camera angles, camera moves and lighting effects. An existing window would allow for filtered light, and the light could have a western barroom look through the addition of smoke in the air—actually, incense burned in a bee smoker.

The size of the barroom suggested that there could be additional patrons off-camera during scenes. The budget limited the number of on-camera characters, so I planned to suggest unseen characters by adding supplemental sound effects: the clank of glasses and bottles in the bar, the tinkle of music on a honkytonk piano, the hubbub of crowd voices. Throughout the pre-production process, a continually revised split script kept track of these ideas as they were noted on its pages. Constant reviews of these additions then allowed producer and director to do the planning that was required to include the ideas in the final production.

KEEPING NOTES

I have found that the pre-production phase involves the generation of many ideas at once. The key to being able to add these elements to a production and increase its richness lies in organizing those ideas in a manageable fashion. While working with a small crew on a limited time frame, a director must be able to track what is necessary to execute those ideas without becoming overwhelmed by details to the point that he loses sight of the overall flow of the program.

By using a split-format script on a word-processing program, a director can regularly modify a script to include new production details. My own approach is to first annotate the script in pencil with ideas as they come to me. They may relate to casting (for example, how old are the bar patrons in relation to one another?) or they may relate to costuming, props or staging elements (banker has a vest and watch fob, newspaper editor wears a derby, etc.).

Then I will go back through the script on successive readings and test whether the details that I have added should stay in: are they feasible, and are they consistent with the writer's intent and with other production details?

Once a detail is clearly seen to be a good addition, and the script is relatively final in other aspects, I will add that detail into the split script that I am working with. This is not the traditional method. It only works when the number of people working on the production is small and the director is in close communication with all crew members. It requires that the director keep track of who has what version of the script and sees that appropriate persons receive a finalized script in time to execute their respective jobs. At some point, the director must sign off on which additions he feels are essential and which are of a lesser priority.

This method does not lend itself to the more traditional large-crew approach, when tasks are delegated early in the process. However, it is a simple approach for the limited time-frame production with a limited crew and ensures that the great ideas that occur at various points in pre-production aren't lost by the time tape is rolling.

The Split–format Script as a Pre-Production Tool

On a smaller-scale production, the inclusion of several key elements in the script format will begin to take it into a form that is quite different from traditional film shooting scripts. Many of the elements that finally are included in the shooting script appear more traditionally on separate lists and scripts that are kept by assistant directors, prop persons, wardrobe persons and sound recordists. But in this approach they are consolidated into one script.

For example, prior to shooting, the split–format script in Figure 3.2 may begin to look more like the one in Figure 3.3.

Shot Numbers

In the new script format, I have included three key elements in my added notes. Shot numbers are indicated to call out the intended breakdown into discrete takes that will be edited together. Since these are each representative of the same basic scene but require different camera placements they are included as parts of a sequence named "101," with "A," "B" and "C" being added to differentiate which of each will be shot from what angle.

Many texts are available on the subject of script and scene breakdown. Those oriented to the director's script study are helpful for initial readings and pre-visualization of scripts. Those that are written from a script supervisor's viewpoint are helpful for developing one's own basic system or embellishing on proven systems. Recommended texts are listed in the Appendix "Resources" at the end of this book.

Figure 3.3: Split-Format Script with Key Elements Added

Wide shot from saloon doors to interior of saloon. 101A: _ _ _ _ _ _ _ _ _ _ _ _ _ _ Includes all 5 characters. Barboy, sweeping, moves across frame with BROOM. SFX: Glasses _ _ _ _ _ _ _ Crowd _ _ _ _ _ _ _	**RANCHER:** You know what really gets me? (Finishes off DRINK)
Two shot Rancher and Bartender. 101B: _ _ _ _ _ _ _ _ _ _ _ _ _	**BARTENDER:** Shhh. Keep your voice down. (As he polishes bar top w/ RAG)
(dolly to include Newpaper Editor)	**RANCHER:** I moved to this territory 'cuz I figured I'd have more choices out here, you know, more options. But it's the same here as everywhere else. **NEWSPAPER EDITOR:** Tell me about it. I hear the same thing over and over. (Also finishes DRINK)
Close-up Bartender 101C:	**BARTENDER:** What about you, Frank? You've done business with some of them other fellas.

Time Code

Note that each shot number is followed by a blank line. This will allow that shot's time code to be filled in prior to the edit session. For example, 101A may involve seven takes. As director, I will indicate Take 7 as my choice. The time code for Take 7 (e.g., T7–1:42:27:—one hour, 42 minutes and 27 seconds would approximate the beginning of the action of Take 7) would then be filled in in the blank space, so that a simple glance at the filled-in script would allow me to cue up the chosen segment of tape during the edit session. Noting T7 lets me know that there may be usable earlier takes if, on second viewing, I am not happy with Take 7.

Sound Elements

In a similar fashion, the term "SFX" is added to the script the first time that a particular sound effect would be required. Whenever that effect was recorded onto time-coded tape, its time code would be entered into the script. Accordingly, the edit for the audio mix track would proceed very rapidly because the editor and I, or I alone, could find each effect as soon as it was needed without seeking additional cross-references.

Having the time code listed on one master director's script allows me to edit the first cut of a program without referring to (read, "searching through") the original camera notes, a process that greatly slows down any edit session and takes the fun and some of the creativity out of the editorial process. If it is too cumbersome to fill in the selected takes during the production, I have found it worthwhile to review the camera notes and fill in the master script with this information prior to the edit session. The one or two hours of bookkeeping result in a much more enjoyable edit session.

Enhancing Script Reviews

Fortunately, word-processing systems usually allow a search of a script for strings of letters such as "SFX." Performing one search of the script for the SFX string prior to shooting will allow the director to quickly create a separate list of all the sound effects that he wants. That list can then be used to search a sound effects library (something very quick to do now that many are available on CDs) and isolate which ones already exist, and which ones will have to be created or captured on location.

Also note that key props such as the bartender's rag, the rancher's and newsman's drinks and the barboy's broom are called out in capital letters: RAG, DRINK, BROOM. This allows a similar quick scan of the script to cull out what props are required that might otherwise be overlooked in a script reading. Many times a director's good idea for a bit of "business" is lost simply because the props were not acquired in time.

CONCLUSION: BENEFITS TO THE DIRECTOR

The approach described in this chapter allows a director to leverage his time and add some of the production values that are otherwise unavailable to anyone who must work with limited time and resources. Some productions developed with these methods have extremely rich audio and video textures, even though they were produced on a limited budget. The key to this approach is to identify all the essential elements *early* in the pre-production process. The ability to do this is developed over time, with practice by the director who makes a conscious effort to imagine the finished program in all its detail.

This kind of organization will allow a director to work more efficiently in all production phases and can often make up for the lack of assistance that a director is faced with on smaller-scale projects. The specific form of any master directing script can be tailored to a specific director's production and editing style. Often, the word-processing system that is available to the director will be a factor in creating a workable layout that is easily edited or updated. See Resources for a list of recommended word-processing programs for video scripts.

4 Choosing the Production Approach: Single Camera and Multi-Camera

PRODUCTION DESIGN CONSIDERATIONS

Somewhere between the initial script review and the final shooting script, the director must decide how large a crew he needs, who is specifically on that crew, and what equipment will be used during each production day. These decisions are important because they each represent a freelancer's day rate or a specific rental cost per day.

Beware of the urge to hire more people than necessary and to rent extra equipment simply because it "might come in handy." These are two quick ways to ensure a short directing career! Although being prepared is the key to good directing, preparing for any eventuality without regard for cost is a fast route to having too little budget to finish the program in progress. Needless to say, a program that is completed only inside the director's mind is of little use to the client.

A Realistic Approach

There is a middle ground that the responsible director must strike. He must first develop a production approach that utilizes a specific-size crew and a particular equipment list (or production "package"). Then he should review the script in light of his production needs, new ideas and new information, to see when he might need additional personnel or equipment.

Just as it is irresponsible for the director to create situations in which people stand about with little to do, it is also wrong for him to ask too few people to perform tasks that tire them too quickly or lead to dangerous situations. For example, he should not ask a two-person support team to execute dolly shots as well as move basic equipment, set lights and record sound. The crew will soon be too exhausted to listen well and to perform their jobs with the amount of attention to detail that is required. In addition, tired crew members could injure themselves and others.

The director should choose a production approach that allows for these limitations. When he finds that a particular shot is necessary or would add an immense amount of production value to the scene, he should see to it that he has the resources to execute it well and safely.

THE SINGLE-CAMERA PRODUCTION APPROACH

One of the most common production approaches is the use of a single camera and recording deck, or VTR, to capture the action in a specific take, either in the actual sequence in which it occurs or out of sequence. The pieces of action are then put together in the editing process to appear as if they are part of a continuous flow of time across each edit point. The "single-camera" approach lends itself well to the small-crew situation.

The "One–Man Band"

In the simplest single-camera configuration, the director can operate the camera himself and control the start and stop of a VTR's recording cycle. Operating as a virtual "one–man band," the director can control all the technical aspects of the production without interfacing with any other crew member—and he can devote all his outward communication to the talent and client. This has the advantage of providing the fewest distractions for the performers. However, it has many inherent disadvantages for the director.

Disadvantages

As he peers through the camera's viewfinder, the director must concern himself with the composition of the given shot, including focus, iris and camera movement considerations. He must *also* concern himself with the video levels, battery levels and sound levels involved in the recording process, as well as the length of tape left on the loaded reel (i.e., will it be enough for the take in progress?).

With all these technical production details to consider, he will inevitably become distracted during some key points of the talent's performance and be less able to judge whether a retake is required. At a minimum, he will have to review some takes off tape while the talent waits for a verdict on his performance. At the same time that he replays a take for client and talent review, the solo director will have to concern himself with cueing up the tape to the correct spot, stopping it after the take ends, and returning the VTR to the correct "STANDBY" to record mode.

If any distraction occurs, the director risks recording over a good take or beginning a talent take without the tape actually recording. In addition, he will rarely have the time or a free hand to make notes about the take which could assist in the post-production process. Without such notes, he will be condemned to "living his life twice," that is, having to review all his footage to see what was usable and what was not and correlating those decisions with the footage after the fact. This is why technicians, VTR operators, or engineers are helpful for more than simply technical assistance.

The Two-Person Operation

One of the most convenient ways to match people skills with production needs is to split the duties of the cameraman/director and the VTR operator/soundman. In this production design, the director sees the image as it is recorded and is responsible for the overall composition, but he is freed from concerns regarding the sound levels and the purely technical aspects of the video signal being recorded. The technician/VTR operator has a monitor which "sees" the same image through the VTR and he can monitor sound at the VTR via headphones.

The director no longer has to worry about the particular mode of the VTR or the place on tape after a review of a past take; the technician will ensure that no good takes are "eaten" or "burned" by a new recording. The technician is also able to listen to each tape on headphones with no audio distractions and to determine when the sound quality is usable or when a take should be stopped due to background noise.

Further, in a two-person configuration, the director's decisions regarding the quality of specific takes can be noted by the technician along with the SMPTE time code or reel numbers that correlate to each take. In the post-production process, the director will not have to review any takes that were deemed useless.

Occasionally, the director can ask the technician whether there are visual distractions or visual problems that he might have missed. Having this second pair of eyes and ears greatly enhances the director's ability to attend to other details without worrying that he might be overlooking something that might haunt him in the edit process.

The Gaffer-D/P-Videographer-Cinematographer

The next most helpful addition to the technical crew is someone whose primary concern is the maintenance of a good "look" to every image before the camera. Good lighting for the video camera can involve constant re-adjustments as scenes change slightly, the sun moves and background objects change. On a small crew, the gaffer's job (a gaffer is a lighting technician), in consort with the video technician, is to see that a consistency of levels is maintained from scene to scene and that each individual scene has a lighting balance that complements the composition of subjects within the frame.

To perform the lighting task well requires constant attention to details as shots progress. It also requires constant physical effort—moving lights, adjusting scrims and barn doors, getting accessories, laying additional power cables, blocking out unwanted sources of light, etc. The small changes that make a major difference in the final look of a program are often the first casualty of an overworked director/ cameraman. I have found that the addition of a lighting expert to the crew can pay off quickly in the overall visual consistency of a program.

Choosing to "Operate" or Not

Many directors like to operate their own cameras, while others are only too glad to give this task to someone else. I have found that there is an advantage to being able to change roles according to a given situation when I am working with a small crew.

If I have a script with demanding dramatic requirements, I will choose to view the technician's monitor and let the camera/lighting person worry about aspects of camera operation that I might find distracting to pure talent direction. However, if I am working with inexperienced talent that I must direct, and they must relate directly to camera, I may choose to be the camera operator simply to provide a greater sense of closeness between performer and camera.

The people in front of the camera may find it easier to relate to me than to the camera operator, since I am giving them directions and concerning myself with their comfort and "presence." This added closeness to me as the camera operator can result in added closeness to the camera lens and, ultimately, in added closeness to the audience. This approach is especially helpful with people who might have a difficult time maintaining good eye contact with the camera lens.

By relating to the talent from directly behind the eyepiece, and consciously not leaving my position behind the eyepiece, I can force them to look into the lens. Otherwise the lens is simply a cold piece of round glass with no personality.

Production Assistants and Script Supervisors

One of the most aggravating aspects of working in video is that one is still a slave to paper! I must constantly use a written script on most programs and must be able to tell everyone where we are in relation to the script, what we're doing next, and how all these shots relate to each other. Often, I spend valuable time finding my own place in the script so that I can keep the production on track and on time.

A most valuable addition to any crew is someone whose primary concern is keeping tabs on everyone's place in the script. This is the role of a script supervisor (sometimes called "continuity") or a production assistant who can be trained to handle this task.

This support is particularly helpful when a director is working in a remote location, or when his hands are occupied with the camera or when he is demonstrating motion to the talent. I find that the "second brain" aspect of a good script person can assist me in sorting out complicated and confusing production details that might otherwise lead to embarrassing mistakes that would have to be dealt with in post-production.

By having one person who is dedicated to making sense out of what is being shot and how it relates to the overall script, I am freed from having to be a constant notekeeper. Further, that person can then take the production notes, allowing the technician or tape operator to work more quickly at the start of each take and in re-cueing takes, when necessary, for review. In addition, the script person's notes are invariably more legible and detailed, and infinitely more informative, than those kept by an engineer or technician, since they have other concerns which distract from content and script considerations.

Additional Personnel

In the single-camera approach, the people cited above play the key roles in allowing a director to attend to the business of directing. Additional personnel could include a sound operator, to be added as soon as a production expands beyond the "talking head" with lavalier microphone style. Lighting and grip support should be added as the visual design and production timetable become more demanding. The key to selecting appropriate crew members is to be fully aware of the details involved in staging each setup in a production, including how long will be required for each setup and how much time can be saved with the addition of each new crew member.

A director soon learns to communicate clearly with his crew regarding the scope of setups and his expected production timetable. This gives crew members the opportunity to tell the director when they feel that he has created unrealistic situations. This also gives them the chance to suggest additional crew or a longer production timetable or a fall-back plan.

SINGLE VS. MULTI-CAMERA PRODUCTION APPROACHES

The most common production approach for video programs involves the use of a single camera and videotape recorder to record the same action from several angles. Each time the camera is moved to a new angle, the action, or at least a part of the action, is repeated. By editing the individual takes from these several angles together as if they occurred in the same real-time interval, the director is able to create an interesting program with visual variety. Essentially, he creates the illusion that several cameras were present at the time of the original action.

In the past, this technique was often referred to as film-style production. Single-camera productions necessitate many edits to create a final program. Videotape was not originally used for programs requiring editing and was chosen over film only for programs which would benefit from "live-switching" of cameras in a studio. These distinctions are now misleading. Video editing is extremely easy to do. Film crews regularly record several cameras at once so that the effect of live-switching can be captured in the film edit process.

However, the single-camera versus multi-camera production decision is one which faces a director on many projects, particularly those which approximate true "screenplay" form. That is, for programs that involve a stage or interior setting and live performance in large blocks of real time, a director may find that the economics of multiple-camera production with minimal editing are comparable to those of single-camera production with extensive editing. Additional factors should then be considered when selecting the production approach.

Advantages

There are two basic approaches that can be used in multi-camera production: *live-switching,* in which only one VTR records the camera choices as they are made on site or in studio; and *isolated* recording, in which several or all cameras' outputs are recorded by individual VTRs so that the final switching decisions can be made later as editorial decisions.

Live-switching greatly reduces the post-production task, since there are few post-production decisions that can be made once the event has been recorded. With only one recording, edits can reduce the overall length of the program or add titles, but no significant changes of angles can be made. Naturally, this reduces the time and cost of post-production.

Both multi-camera approaches allow the director to preview shots before switching them. He does this by comparing images on several monitors in the control room or truck, which means that he will not be surprised later when two shots are joined with an edit. (In the single-camera production, shots are staged out of sequence and the director must keep good visual-continuity notes so that he does not create jump-cuts or inconsistencies in real-time action.)

Continuous Time

But the most important advantage to using a multi-camera approach is that it allows the action to flow in real time and in the actual sequence of the final program. In other words, if a character in a script enters a room, greets his wife, learns that his son has been kidnapped and then phones the police, the sequence in which these individual actions are taped will be the same sequence in which they appear in the final program. The changes in character as the action unfolds will occur in the sequence that the actor experiences them.

This fact—that the actor experiences events in the actual sequence that the character experiences them—allows the director and actor more freedom in the portrayal of that character. Since each take "starts from the top," slight modifications to the delivery of a given line or to the actor's movement will flow naturally into succeeding action and dialogue. If parts of the scene were to be shot out of sequence (as in single-camera production), these modifications might present problems.

Out-of-Sequence Staging

In out-of-sequence staging, the entire change in character development must be mapped out in advance so that the talent's action and performance level is consistent with what precedes and succeeds in the final edit. When taping with a single camera, a conscious choice is made to trade off the inconvenience of this restriction with the convenience of shooting all shots from a given angle at once. For example, in the sequence described above, the entrance and final phone call may call for wide shots. Consequently, these two parts of the scene would be shot first. Then, the closer shots, which encompass the intervening action, would be taped.

When the closer shots are taped, the director must take care that the talent's performance and action will cut together with the preceding entrance and the following phone call. This is not always as simple as it might initially seem. Let's say the actor discovers a more interesting way in which to deliver lines, or a new bit of action with props or other characters for the "middle shots." He and the director will be restricted by the need to make these additions only insofar as they remain consistent with the opening and closing shots. Thus, the director must closely examine a script to see if it will benefit from multi-camera or single-camera production; that is, from in-sequence or out-of-sequence staging.

Script and Staging Considerations

If a script involves rapid interchange between characters, the director may lean toward a multi-camera approach since the dialogue will be recorded as a continuous track with camera cuts or visual edits being a separate element. In other words, an additional advantage of multi-camera, in-sequence staging of a scene is that a minimum of audio editing is required afterwards.

For the same reason, this approach allows for "overtalk," or overlapping dialogue. Actors do not have to avoid speaking before their fellow actors finish their own lines. This will lead to more natural dialogue than if actors have to help the editor by avoiding overtalk, as they must if audio and video edits will occur later.

There are many advantages to multi-camera production in terms of spontaneous performance, but the director must prepare thoroughly to take advantage of them. Since more cameras, crew and equipment mean more hourly expense, the director must be able to put more script on tape per hour to compensate, if he is to stay within budget.

For location productions, the additional equipment will require more setup time and more time to wrap. Additional cameras will require additional time for matching, and more cables will have to be strung for communications and monitoring. In conjunction with a larger crew, this will also entail more disruption at the location. If the production is using a client's offices as the location, this may be a major consideration. In addition, the greater amount of equipment will increase the chances that there will be delays if any equipment fails.

COMPARING THE TWO APPROACHES

In the single-camera production, the director can accomplish his setups more quickly with less disruption, but he must be able to guide each step in his out-of-sequence staging very carefully to accomplish a smooth finished program flow. If time and budget allow, a complete prior rehearsal of the script from start to finish will aid greatly in achieving this consistency. If this is not possible, then he should at least have a full rehearsal on the set or the location of each major scene prior to taping.

The advantages that accrue from single-camera production include a more varied framing of shots. In multi-camera production, as the action unfolds, the director can only call for shots that can be set up quickly while a camera is off-line (not being used to record). This usually means a minimum of height adjustment and limited camera movement since these setups are accomplished in real time, and the director can usually leave a camera unused for only a short time. With a single camera, the director is able to take more time framing each shot and this results in a more varied and usually more "designed" visual style.

For example, a low angle can be used to accentuate the subject's power or stature. Higher angles can be used for the reverse effect. Foreground and background objects can be placed precisely prior to rolling tape on a given scene. But these effects can only be achieved when there is enough setup time for each shot. During multi-camera production, the director is usually occupied with several real-time tasks and can spare little time to frame shots precisely.

For a multi-camera director, the timing of his camera cuts (calling the shots), the flow of the dialogue and blocking take up much of his attention. He must rely on the camera operator's judgment for much of this task and can only refine or trim shots to a limited degree. In the single-camera approach, he can collaborate more with the cameraman, spending more time creating the look of each shot in the program, often setting the framing exactly to a pre-visualized shot.

Differences in Preparation

When a director chooses one method over another, a different emphasis is placed on the pre-production task. If a single-camera approach is chosen for the greater control over individual shots that it affords, the director will have to spend more time in pre-production visualizing the overall program, but can spend less time defining the individual shots. This approach will allow him greater latitude and the time to refine his decisions on the set or location.

In adopting the multi-camera approach, the director allows himself more freedom to experiment with overall program flow on the set or location. Once production begins, however, he must be prepared to make his decisions quickly and decisively. He also must define his shots in terms of blocks of multi-camera placements—a different sort of shot list than he would prepare for a single-camera production.

Aside from these basic differences, many of the pre-production tasks remain identical for either production approach. One of the first and most essential tasks is the selection of talent, the subject of the next chapter.

5 Selecting Talent

THE IMPORTANCE OF CASTING

The most critical element in virtually any video production is the performance of the people appearing in the production. There are a great number of factors which affect how well talent appear to the audience when they perform on camera: how often they have practiced in front of a camera, how comfortable they are with themselves, how much public speaking they might have done in the past, how well they know their material or have memorized their lines, and how much native ability they have.

The extent to which a director can affect the overall performance of any given actor or non-actor is directly affected by what he or she brings to the situation. To the degree that the director can spend time with the talent, the director can shape a performance by suggesting characterization, shading a delivery, coaching gestures, suggesting timing and blocking changes. But these will always be relatively minor levels of change when compared to the basic background and capabilities that on-camera talent bring to the production.

Most theatrical productions provide not only weeks of rehearsal, but nightly performances that can be reviewed, analyzed and refined. Feature films often provide weeks of off-camera rehearsal time that allow the talent and the director to shape character interactions. Newscasters appear on-air daily and often twice a day. This gives them and their producers and directors ample opportunity to develop highly polished performances.

Since any audience is exposed to polished performances regularly on the news, at the movies and in the theater, it expects a high standard of performance quality. The performance of your talent will always be judged against that of these highly professional, practiced and polished performers.

Recognizing this, the director should treat the selection of talent with the utmost care. An ineffective actor or actress can create a frustrating situation in which an incredible amount of directorial effort can be spent to little recognizable good. Increasing a 10% effective performance to a 20% effective performance may be a huge feat on the director's part, but the audience or client will simply perceive a poor performance, compared to what they have come to expect from the medium. They perceive the overall program as one piece and cannot know how bad it might have been.

Organize the Process

Those who haven't much contact with actors and actresses often think of them as a breed apart. Some people think that there is a mysterious way to communicate with actors and actresses and that selecting the right talent for a production is a mystical process, that the director must possess a special intuition to know who will perform well in front of the camera.

The fact is, successful talent selection is only partly intuitive. Most successful directors use a systematic approach (whether or not they are conscious of the fact).

The talent selection process can be broken down into the six steps below. The first three are covered in this chapter, and the last three are discussed in the next chapter.

1. Developing a knowledge of the available talent pool.
2. Reviewing the production demands on the talent.
3. Making the talent request.
4. Preparing for the audition.
5. Conducting the audition and callbacks.
6. Making the selection.

KNOW THE "TALENT POOL"

Every director must work with two basic limitations in regard to who might be used as talent for any given production. First, the production will take place in a given geographic area and at a specific time. This will determine who is available to work on the production. Second, the production has a set amount of money for talent fees. This will limit who, among the available talent pool, can be expected to choose to work on the production.

The director should treat each talent selection as an opportunity to gain a better sense of who's who in the local talent pool. This will make the next selection a little bit easier as he adds to his knowledge base. Each talent review, audition and selection can be seen as part of the director's education in his craft.

Talent Composites

A key tool for learning about the local talent pool is a talent file. The basic building block for that file is the composite that every serious actor or actress carries to auditions, has on file at agencies and sends to anyone who might be interested in casting them. The composite, or "comp," is typically composed of an 8½-in. x 11-in. sheet. One side has a full close-up head shot, and the other side will usually have four to six smaller photographs of the talent in various poses to show a range of types that he or she might play.

Usually, the talent's composite is black and white with hair color, height, weight, clothing sizes and age-range noted in print. (This is the same format that is utilized by modeling agencies.) Since most scripts require that a performer look the part as well as have the ability to portray a character, the composite is a good way to begin to gauge a talent's basic suitability for a role. However, since actors and actresses can change their appearances radically through makeup, wardrobe and movement on camera, the still photographs rarely tell the whole story.

Composites serve best as basic reminders to a director, a shorthand for remembering who's who in the talent pool. Any serious director should develop a file of composites on every actor and actress that he interviews, meets, auditions, or sees in other works. He should note on their composites in what productions they have performed, how they performed, what their strengths or weaknesses were, and any personal observations.

The good director will develop a keen sense for the talent's unique characteristics. He will note even the smallest details so that he can refer to specific actors and actresses long after seeing them. This personal sense for the talent will greatly help the director with future casting decisions.

Using Composites with Clients and Talent Agencies

Composites can also serve as a common language device when working with a client. If a client visualizes a spokesperson or model salesperson, the director will be able to guide the client as the client reacts to specific talent photos, indicating how the particular actor or actress might handle the role under consideration. This may prove essential to controlling the casting process—since some clients can become insistent on casting talent based solely on their looks.

After the director understands the client's needs, he should pre-select talent photos or composites based on his own judgment, making strong initial recommendations to the client. Often the fewer the choices presented to the client, the better the selection. Basically, the director should exercise his skill and experience at this point and argue for the best talent. After all, he will have to work with them and be responsible for their final performances.

The director should review his talent files after reading a script and prior to speaking with a talent agent. This can save an immense amount of time, because it allows the director to request that actors similar to a specific actor be scheduled for an audition. Or, for example, he can tell an agent that he needs someone with the age and basic look of one actress but with the added sophistication of another. Since human qualities are difficult to put into words, this allows an agency or casting service to work with the director to zero in on the right "type" and eliminates wasted effort.

Resumes

Typically, actors and actresses will attach a resume to their composites. The resume will include their past acting roles (usually broken down as to stage or film), as well as any applicable schooling, special classes, lessons or workshops in which they have participated. A lot can be learned from carefully reading a talent's resume, as we show below.

Stage versus Screen Experience

If the resume includes only stage roles, there is a strong chance that the actor or actress has had limited or no experience before the camera. This should be noted by the director as an area to explore with the talent, since stage performance does not always translate easily to the screen.

For example, there is a great difference between the levels of voice projection required to address an entire live audience from front row to back in a theater versus the voice requirements for a recording to be played back on a television speaker. Some voice characteristics which work quite well in the theater can be thoroughly aggravating when recorded with the close presence of the microphone. At the same time, the microphone's sensitivity provides an opportunity for much greater nuance in inflection.

So, too, the camera, which allows close-ups, provides for much subtler shades of body language. Techniques that utilize the physical space provided by a static stage before an audience are quite dissimilar to those that work in an extreme tight shot. And, in fact, the same performance that appears dramatically genuine on stage can appear false and "stagey" when captured by the lens.

This is not to say that the director should automatically eliminate experienced stage performers from consideration for film or video roles; rather, it suggests that the director should be prepared to test their camera awareness during auditions.

Models versus Actors and Actresses

A resume that includes no stage background but does include several feature films, without indicating specific roles might mean that the talent may be more experienced as a model than as an actor or actress. They may have appeared as extras or "atmospherics" on a set because they lent the appropriate look—not because they could deliver lines.

If actors or actresses have had speaking parts, they are usually quick to indicate it on their resumes. If a role is not indicated by a character name, it often means that the role was that of an extra. At the very least, the director should probe for what the role encompassed.

Commercials versus Longer-Format Productions

A resume that indicates primarily commercial work may not demonstrate that the talent can sustain a role, or even memorize lines. Many commercials include few, if any, spoken lines and very little character development. They are designed to give the audience a quick "read" of a character. The talent who can do this in the short format of a 30- or 60-second commercial doesn't necessarily have the skill to sustain a characterization through 12 minutes of a corporate video or 87 minutes of a feature film.

Overall, the director should proceed with the selection process by trying to find the best base of already-established talent skills on which to build and shape the performance of the production. Knowledge of the talent pool often shapes the next step in the process: reviewing the script to define the talent demands.

GAUGING PRODUCTION DEMANDS ON THE TALENT

Every script will have varying types of talent demands. Assessing these demands and casting only those actors and actresses who can meet them is one of the practical skills that a successful director must develop. This process is not always completed upon a simple reading of the script. It requires the director to review the specifics of his production design, production schedule, size of crew, crew responsibilities and client involvement.

The director must ask himself a host of questions, including:

- How much time with the talent will I have for rehearsals?
- Will there be time (and budget) for rehearsals prior to the shoot, or will rehearsals have to be done during setup time on the day of the shoot and in between setups?

- Will there be crew members who can assist with script review with the talent when I am not available?
- Will the talent have to master specific skills (such as operating machinery) as well as learn their lines?
- Are specific prompting devices required, including cue cards, teleprompters, or ear prompters?
- Will scenes be shot out of sequence—and if so, is there complex character development involved in any given role? How great will the distractions be?
- Will there likely be script changes on site?
- Will the talent have to speak over loud background sounds such as those in a data processing center or under a helicopter?
- Will I be able to work in close physical proximity to the talent (as single-camera director), or from a booth, control room or truck (as a multi-camera director)?

All these questions can be pertinent to production demands on talent and director. The director should have a clear vision of all the distractions with which he must deal, so that he can project how much time he will have to shape performance and how proven an actor or actress he must cast.

"Old Reliables" versus New Faces

The smaller the crew, the greater the demands on the director to multi-task and the less his ability to focus on the talent's performance in isolation. As a result he may place a premium on self-directed, accomplished actors and actresses who can maintain a consistent performance without constant coaching. These performers tend to be highly used, even overused.

However, the trade-off is that these proven performers may look good because they consistently act within a safe, narrow performance range. The director who wants to make the most of any production will always review the script and production design and then cast for the most interesting performance possible without jeopardizing the production by taking too great a risk with unknown factors. Basically, the ambitious director will define a "safe tolerance" zone of performance range for the talent, and then push to the limits of that zone.

More specifically, the director may find a very interesting actor who might bring a unique characterization to a role, but who doesn't make the grade at the audition. The director will have to judge to what extent he can "bring the actor along" to the required performance level. Here is where he must have done his homework prior to the audition. He must be able to assess whether the actor can learn quickly enough, given the limitations of the production design and schedule, to not risk the overall success of the production.

Newscasters versus Actors

It is common for newscasters in small and mid-size metropolitan areas to be cast as corporate spokespersons. However, though reporting the news and speaking on camera as a spokesperson have overlapping techniques, they encompass different skills as well.

The newscaster who has done primarily standups and anchor work, in which he has remained stationary, can lose his self-confidence when confronted with the body movement that must be incorporated into a spoken part. He may not have developed the ability to hit exact marks during movement with dialogue, or to relate to a moving camera (and a moving teleprompter).

The director should note requirements such as these when preparing for auditions in which stage persons might transition to on-camera roles, newspersons to spokespersons or dramatic roles, or dramatic and character actors and actresses to spokesperson roles for the first time. Many of them may be quick to learn, with no risk to the program, but those who are slow to learn new skills could jeopardize an entire production.

Teleprompter Requirements

Teleprompters are in common usage on productions that involve long or technical copy, and there are particular skills which actors acquire to mask the fact that they are reading. These include head movement and regular glances off-camera, as well as body positioning and gesturing while reading. Inexperienced teleprompter readers will often stare at a single spot, producing a vague, glazed look, or in some cases, seem to track the copy from side to side as they read each line.

If a teleprompter is required, and a certain subject-to-camera distance is needed for a shot, the director must assess whether the talent can, in fact, read the script from the teleprompter at the required distance. Too often, eager actors and actresses will be overly optimistic about their abilities. (Vanity often affects their judgment, as well.) It is the director's task to define and test the critical performance requirements of the production being cast and to prepare for the audition or talent call in a fashion that will unearth any problem areas.

Memorization Requirements

The reverse of this problem may occur when a director casts an experienced teleprompter reader in a role that requires some degree of memorization. The news anchor, for example, who used to do memorized standup pieces as a field reporter, but who currently does in-studio news reads, may have lost the ability to memorize and not be aware of this fact.

The director should always review the program's script and note the relative amounts of voice-over copy (no memorization, no prompter), on-camera copy with prompter or cue cards, and on-camera copy without prompting options, and discuss the amount of required memorization with the talent. (It may help to color code each type of requirement for easy reference.)

Terminology Demands

If a script contains technical jargon and words difficult to pronounce, the director needs to develop a test script. An otherwise perfect performance can be completely undercut by incorrect delivery of industry-specific terminology.

Doctors will spot an actor playing a doctor if the actor doesn't use the appropriate terminology in the same offhand manner that a doctor might. Engineers react to a too-careful pronunciation of acronyms as a false performance. The director must test for the range of the talent's ability to incorporate the buzzwords, slang and technical verbiage into his own spoken vocabulary in a truly believable manner. Otherwise, the sense of reality of any given scene could be broken.

New Technology: The Ear Prompter

The "ear-prompter" is a small tape recorder into which the talent reads the copy as the director shapes inflection and pacing. The tape is then rewound, the recorder placed inconspicuously in the talent's clothing, and the script is played back through a hidden earpiece (which may be wired to the recorder or can be transmitted in wireless fashion through an induction coil around the talent's neck). The talent simply hears the words he is to speak a few seconds before he speaks them.

The ear prompter can save countless retakes when a traditional prompter cannot be used. The talent can simply re-record passages as required to incorporate any on-site changes. Many directors are taking advantage of this technology and are seeking on-camera performers who can use this device.

WORKING WITH THE AGENCY

In most metropolitan areas, actors and actresses use the services of talent agencies and talent agents to represent them and to find job opportunities for them. The agencies' clients are producers, directors, talent directors and casting agents. The agencies typically request a 10% fee for their services. Usually, the director deals with the agency rather than the talent in the initial phases of the casting process.

Agencies can be of great help to the director who knows how to utilize them effectively. To do so, the director should, after script review, draw up a list of the characters he needs to cast. Included in that list should be the "type" requirements, if any, that are necessary for each role. These would include male or female, age range, height and weight, and any particular character features or aspects that play prominently in the characterization.

Figure 5.1 is a sample character description for a fictitious casting session.

Know the Agency's Agenda

The character description provides only the "type" requirements dictated by specific script needs and doesn't reflect what the actor and actress might bring to the interpretation of the roles. But it does serve as a starting point to limit the talent call.

This limitation is necessary in light of the fact that the agency is motivated to get as many of the talent they represent to as many auditions as possible. This serves to make the actors feel that the agency is doing a good job for them by creating

Figure 5.1: Sample Character Description

(to be sent to a talent agency)

Yoshi

A 30- to 35-year-old Oriental male computer programmer (preferably Japanese), with an athletic appearance in casual work clothes—i.e., cannot be noticeably overweight or bald and must be able to perform without glasses. Some comic ability required. (Ability to touch-type would be a definite plus but not an absolute requirement.)

Maxine

A 40- to 45-year-old blond female executive who grew up in Brooklyn and retains her accent. Must look at home in a tailored suit. Slightly overweight. (Similar to Nancy Randall [represented by the same agency], but with a somewhat older appearance.) Ability to drive a stick-shift sports car a must.

DATES OF PERFORMANCE:_____

DIRECTOR:_____ PHONE:_____

work opportunities. However, it is a nuisance to a director with limited time for casting, especially if many of those who appear for an audition are simply inappropriate for the parts.

Sometimes an agency may send out a core group of people to many auditions unless a director specifies otherwise. That core group will consist of people who present themselves well in any audition. The agency is confident that they will represent the agency well and hopeful that they may get the parts, even if they are not within the specifications initially given. If the director recognizes this pattern, he may tell the agent that he doesn't want to see persons A, B or C at this audition.

Inform the Agency

The agency should always be told of any scheduled or tentative production dates so that it can let the director know of any potential conflicts for actors. It is an obvious waste of time for a director to cast someone who cannot make a firm production date.

The agency should also be advised of the rate of compensation. In a union (SAG/AFTRA) environment (see Resources for the national address), this is clearly defined in the union contract as "scale." This is the standard minimum rate for an on-camera day-player (or spokesperson) in a commercial or in a Category I or Category II industrial production (Category II refers to point-of-sale video or film work).

If the budget allows for compensation only at scale rates, the agency must know this, since it may represent talent who can command double or triple scale due to their popularity. To cast such persons only to find out that the budget can't afford them is disappointing and a waste of time. (If the client has seen a candidate who turns out to be unaffordable, this can lead to a nagging feeling that the production could have been better than the director made it.)

A Caveat About Non-Union Talent

On the other end of the scale, the director should beware of situations in which the client dictates the use of non-union talent for budget reasons. In most metropolitan areas, the better, more experienced actors and actresses have had the opportunity to perform in union situations and, as a result, have joined SAG or AFTRA.

This means that there is a greater risk in casting from the non-union talent pool. The net savings of a smaller day rate versus the need to do many retakes (which can easily lead to overtime for cast and crew) may be a negative savings. It may well cost more, on balance, the utilize the less-experienced talent.

Basically, the less proven the talent is, the more time the director should devote to the audition process. In no instance should a director agree to utilizing unproven talent without extensive auditioning. Sometimes a certain quality (familiarity with complex content, appropriate visual appearance, etc.) dictates taking a risk with unproven talent. However, the director should always keep in mind that a convincing performance often makes or breaks a production, and should take the necessary steps to ensure that the talent can deliver a convincing on-camera performance.

NONPROFESSIONAL TALENT SELECTION

Many times, companies with limited budgets or specialized needs request or demand that the talent for a production be drawn from their own personnel. Often, the name, background and expertise of the selected person is intended to lend credibility to a production. The director should act as the judge of the person's potential on-camera performance: is it adequate to lend the requested credibility? Can the person's performance be brought up to the necessary level?

If an on-camera presenter has distracting speech habits, exudes tension, or is hesitant and lacking in confidence, it will undermine the desired intent. The client should be discreetly informed and presented with alternative approaches.

One of these approaches would be to carefully audition professional spokespersons to find a good match to the company's image. Another would be to break out a script into smaller on-camera segments, allowing the presenter to become a narrative voice over visuals other than himself.

The director should be realistic about the level to which a nonprofessional's performance can be improved and insist on adequate rehearsal time to actually make an impact on the presenter's performance.

Cost and Cooperation Considerations

If the client is recommending that an executive of the company present material that could be more effectively presented by a professional spokesperson, and budget considerations are driving the choice, the director may wish to address the problem in terms of real costs. He might suggest that the cost of the executive's rehearsal and performance time could easily exceed the going rate for an on-camera professional spokesperson and wouldn't guarantee an adequate performance.

If an executive is to appear in a leadership role on camera, the director should arrange an early meeting with him to determine the level of cooperation that he will provide. If it is minimal, meaning little or no rehearsal time, the director may wish to ask the executive whether any other alternatives have been considered. Perhaps another officer of the company is more suited to the performance and may be considered appropriate for the task. The first executive may welcome the opportunity to be released from the obligation, a fact that no one in the company had considered.

A Final Note of Caution

When company personnel are asked to fill dramatic or comedic roles, the director should particularly insist on auditions to gauge the skills of the suggested participants. This can be more relaxed than an agency audition. The director can use the opportunity to catch weaknesses and rework the script to better incorporate the available talent effectively.

If such a compromise isn't possible, he should take the initiative to suggest that the existing script be produced using professional talent. If the client insists on using nonprofessionals, the director should suggest that the entire concept be reviewed. The script should then be redeveloped to match the skills of the nonprofessionals. Nothing is more difficult to direct successfully than a dramatic or comedic script acted by inexperienced performers.

(See Chapter 8 for a discussion about directing nonprofessional talent.)

Final Hints

Reviewing one's own talent file and using a person known to both director and agent as a frame of reference, or asking an agent about the skill of someone in the director's talent file, compared to someone else, can speed the initial talent selection process. It is also helpful to prepare a standard set of materials for each audition so that each candidate starts from the same point of preparedness. This is covered in the next chapter.

6 The Audition

PREPARING FOR THE AUDITION

Like many other aspects of production, a successful audition is usually the result of careful preparation: the director should develop a clear sense of what he is looking for in each role. These elements may consist of items such as the ability to develop a comic interplay with another character, the ability to lend an authoritarian or experienced air to a delivery, etc. There may also be subtler nuances that a director seeks as he conducts the audition.

He will be looking for ways in which an actor might alter a role and make it into something more than the script or the director had suggested. In order to recognize these possibilities, the director who conducts his own auditions will want to pay attention to as much detail as possible.

Review the Script

There are several ways to create a more focused audition. The first is thorough script preparation. For each role, the director should zero in on the parts of the script that will best allow an actor to demonstrate his potential character treatment. (The same script passages may not work for all characters. One passage may give a good sense of Characters A and C interacting, while a different passage may give a good sense of the interaction between Characters A and B.)

In the interest of time, and to provide the opportunity for re-readings, the director should create "sub-scripts" tailored for the characters who will read opposite one another at the audition. These will consist of key passages that are long enough to allow some characterization and possible character development, but not so long as to become redundant.

Schedule Appropriate Cross-Readings

Often, characters in a script have key interactions, and the interplay between them is important to the plausibility and enhancement of the script. The director should pick passages that will allow for "cross-readings" between key principal characters and schedule readings accordingly. Three different actors who might play Yoshi, in the example in Chapter 5, might each be given the opportunity to read opposite three different actresses who are being considered for Maxine's part.

This will allow the director to gauge the impact of the interaction of the characters. Noting that the whole is sometimes greater than the sum of its parts, the director may find that a particular actress brings out a more interesting aspect of an actor's performance than was found in a solo audition or opposite a different actress.

Test Talent Redirection

In addition, for each character, the director should decide which passages can be interpreted in more than one way and deliberately test how each person takes redirection. The director must explore not only the interaction among his characters, but also the performers' interaction with him. This will be key to his ability to shape performance at the time of taping or filming. By redirecting the talent he will find out whether there will be a healthy working relationship later in the production process.

In preparation for this, the director should analyze ways in which the talent might initially read a given passage. Then, after the talent's reading, he should ask the talent to re-read the passage with a much different interpretation.

For example, Yoshi might deliver a statement regarding down-time among mainframe computers as if it is a fact everyone knows (virtually a "throw away" line). Then, on a second read, he might be directed to deliver it as if he just discovered a most startling fact, unknown to anyone but himself until this very moment. The same words are delivered in the same context but with a very different intonation and pacing in response to new directorial guidance.

The actor should not be aware on the first read that a second interpretation will be requested, otherwise the actor will tend to overstate one performance aspect so that the difference will be artificially highlighted.

Eliminate Surprises on the Set

Creating this opportunity for redirection will allow the director to discover how well each actor listens, how aware of his own performance each one is, and how well each one can control his performance under direction.

Without a test like this, the director may admire a good reading by a particular actor and cast him. But later the director may find that the actor can only deliver the same reading he gave initially. Testing for this flexibility during auditions will help avoid this problem.

Prepare Character Synopses

Another helpful element in audition preparation is the creation of a synopsis (see Figure 6.1) of each character who will be auditioned. The synopsis can be placed as a cover sheet for each script. It allows the actor to review the basics of the characters, as seen by the director, while he is studying the sample script prior to the audition.

By giving the same initial description to each actor, the director allows everyone to start on the same foot. (The description would vary greatly by the end of a long audition session if the director were to have seen many different interpretations and changed his initial verbal instructions accordingly.) This method also saves a lot of time if many persons are being auditioned and reduces the director's fatigue.

Figure 6.1: Sample Character Synopsis

> **YOSHI:** A second-generation Japanese who grew up in a Los Angeles middle-class suburb. Successful in school, he has had an easy time being a high achiever and this is reflected in his casual manner. Never married, he has enjoyed his single status but is beginning to seek more depth in his relationships.

Let Actors Suggest Blocking

The director should also pay attention to some basic blocking opportunities. What might the talent do with the suggestion of physical movement, how might they amplify or suggest additional bits of "business" that aren't directly called for by the script? This allows the director to explore how each actor might add new dimensions to characters in the physical realm.

The director should let each actor know that such freedom is allowed. If not, the director may be impressed by the few actors who are bold enough to take the initiative themselves without coaxing. (This opportunity also gives the director a form of free advice—actors who aren't finally cast in the role may suggest workable ideas.) A general rule is to keep blocking as simple as possible.

Create Priorities for Casting

As a last step, the director should review all the technical, artistic and "type" demands on the talent in the production and create a clear set of casting priorities for each character. These should be based on his knowledge of the program's objectives, the client's needs and the audience's expectations. Then he can note where he can and cannot compromise. Essentially, he should decide which characters are key to the success of the program so that he uses his audition and selection time to the best advantage.

CONDUCTING THE AUDITION

Time Factors

Auditions are typically conducted under time pressure. This is because most agencies are accustomed to casting for commercial productions, in which decisions are made quickly and producers and casting directors often must cast many parts in a short amount of time. The scheduling of fifteen or more auditions per hour is not unusual in these situations.

The director of a longer-format corporate production often requires more time with each actor, because he may be casting for actors who will have much more time in front of the camera and far fewer possibilities for retakes. It often takes much more audition time to determine whether an actor can perform a given role. There are often more technical, memorization and delivery concerns involved than in the typical commercial in which a quick characterization may be of greater importance. Consequently, the director may have to set the standard, rather than follow the agency's lead, and specify that actors be scheduled at 12- or 15-minute intervals instead of 4-minute intervals.

Mechanics and Logistics

If the audition is conducted at the agency, the director should be conscious of the agency's space limitations. The director should make every effort to stay on schedule so that a "stacking-up" effect doesn't slow down the normal flow of work at the agency. Actors tend to be very outgoing and their interaction can become intrusive in an office environment.

The director should use the agency or an assistant to direct actors to their scripts as they arrive so that he is not interrupted while conducting an audition. Auditions should be conducted in isolation from incoming actors so that succeeding performances are not shaped by prior performances. This allows each performer to approach the audition in as fresh a fashion as possible.

Easing Tension

The audition environment is not a comfortable environment. It is comparable to a job interview situation, with the very real threat of rejection for the talent. The talent may very much want to get the role for which they are auditioning. Further, there may be onlookers besides the director in the room. The director should be sensitive to this fact and, assuming that a more relaxed actor gives a better performance, try to put the talent at ease.

Introducing oneself along with any onlookers and making casual conversation about the talent's composite and resume helps to ease some of the tension that the less-experienced actor feels. It also helps to explain any aspects of the role that may have occurred to the director since composing the cover sheet and it helps to ask the actor if he has any questions regarding the script or the production.

Probing: Resume-Generated Questions

The audition is also the time for the director to seek answers to questions about the actor's resume, e.g.: "What role did you play in *Twelfth Night*?," "Who was the director in this film where you played Staff Sergeant Kehoe?," "How did you like doing a stage play for film?," "Have you done many corporate projects—for whom?," "Have you used a teleprompter or an ear prompter?," etc.

Avoid "Script-Lock"

The director should tell the actor how far into the script he should read and where any particular cues are located. The director should never be caught having to feed lines, that is, to read the part opposite the talent—that is being locked into the script reading himself.

This will divert his attention from the talent as the director finds his place in the script. It also means that the director cannot watch the talent's reactions. Since reactions and nonverbal cues are often a large part of a performance, this means that the director misses much of the characterization that the actor provides in the audition. The director should ensure that another person reads any parts opposite the actor who is auditioning.

Use Videotape

Whenever possible, the audition should be videotaped. This will help the director and any other decision makers judge how the talent's performance translates to camera. There may be quite a different perception of a performance when the talent isn't present. Some actors create a strong impression in person that doesn't hold up when "screen–tested." Others may not be quite so impressive in person, but have a great affinity for the camera.

Reviewing tape at a later time, in a less distracting atmosphere, will provide a much better sense of the talent's on-screen performance. Later, the videotape can serve as a supplement to the director's files of composites and head sheets, so that he can get a good sense of an actor's performance range without having him read for another role. This can save a great deal of pre-production time.

Test for Director/Talent Conflicts

After the talent's initial read, the director should thank him and let him know which aspects were what the director wanted. Then the preplanned redirection should be given, with close attention paid to the response. If an actor takes issue with the director's suggestions, the director should refrain from debating the point (unless the point is a key one and seems to have merit). Instead, the director should move the audition along and not waste time with nuances of interpretation.

Rehearsal—when the director has time to listen—is the appropriate time for actors to make their directorial points. The actor who doesn't recognize this is unlikely to be sensitive to other pressures and cannot be relied upon to exercise the judgment that a busy director requires.

Take Breaks

The director should allow a break after every four to six actors are seen—to discuss details that were noticed, to compare impressions with others doing the audition, to make notes, and generally to switch gears. Otherwise, after several hours of auditions, his ability to discriminate between performers, note details and remember personal impressions suffers.

AFTER THE AUDITION

Review the Choices

Sometimes the final decision for a given role may seem obvious as soon as the last person is seen. Even so, it is advisable to review tape prior to making a final decision. Some initially insignificant habits or traits can become bothersome upon prolonged exposure to the talent on tape. (I once met an actress who had great stage presence but had a pronounced sibilance in her speech pattern. The microphone and recorder combination accentuated this, but the problem did not seem obvious in person or on stage. However, no one could tolerate her recorded voice for any length of time, and no amount of audio processing significantly reduced the problem.)

Reviewing tape the same day as the audition is held reminds the viewer of things that may not have been captured on tape but are in recent memory and these should be noted. A second review the next day is often a good check to validate initial perceptions.

If a great number of candidates have been auditioned, if may be well worth the time to edit down the audition tapes to the few who are possible selections so that they can be quickly judged against one another. In this fashion, their similarities and differences stand out much better.

Allow the Agency to Announce the Casting Decision

If the talent was called through an agency, they should be informed of the casting decision by the agency. This maintains clear roles and allows the director to concentrate on the next steps in the pre-production process.

Most actors would much rather know when they did not get a part, rather than be left in doubt. Many have indicated to me that they would also like to know that they did not get a part because they were not the right "type," rather than because they did a poor reading.

Presenting to Clients

Once a director selects the talent for each role in a production, he should present one selection to the client and not a range of possibilities with one strong favorite. After all, the director is the person who is responsible for the final performance and has the experience on which to make a solid judgment. The client may not be able to discriminate between shades of performance and might react to less pertinent aspects, such as the performer's off-camera personality or wardrobe.

Most clients appreciate the director who is willing to make a strong recommendation and provides the reasoning for it. This serves to build client confidence in the director's judgment and experience.

7 Preparing the Production

INTRODUCTION

Many decisions must be made in the course of shooting any film or video production. They range from which visual will be seen while each segment of copy or narrative is heard to how many camera setups will be done in a given day and in what sequence they will be staged. Some of these decisions can be made during the production preparation step. Many must be made in the course of the shooting process itself.

The director of a small crew is faced with numerous creative and logistical decisions during the shooting day. In the course of directing an actor and a crew, he will constantly be asked, "What are we doing next?" The more quickly he can provide detailed answers, the more effectively he will be able to delegate tasks to his crew. Proper pre-planning will also allow the actors to prepare for the next scene independently.

The effective director can better maintain a view of the big picture when he has prepared a thorough production plan, rehearsed his talent and briefed his crew. This frees him from the small-scale logistic decisions that get in the way of maintaining that overview. It enables him to check whether the talent is performing in a consistent fashion and developing believable characterizations, whether the composition and framing of each shot is complementing the message, or that lighting changes for good dramatic reasons within specific setups.

Thus the director must prepare the production in a fashion which allows him to decide very quickly where he is at any given point in the process, to determine what is expected from the current scene, to recall which scenes are relative to the one being shot, and to know what preparations for future setups should be going on at the moment.

THE DIRECTOR'S TOOLS

A director develops this sense by pre-visualizing the finished program. He records and translates that vision through his script breakout (which will eventually serve as his master script), his shot list and his shooting schedule with its associated production notes. These tools are most effective when they are all created with a thoroughly understood finished program set clearly in his mind.

During the production itself, two sets of notations will be kept. The first notation will be the shot log, or camera notes. This is a continuous log that represents every take that is recorded, including incomplete, or "broken," takes. The shot log will serve as a complete record of decisions made during the production. It represents, in a linear fashion, all that has been recorded during the production.

The second set of notations is the recording of selected takes and possible selections on the director's breakout script. This system allows the director to edit the show quickly from the breakout script itself, referring to other notes only as necessary. For this reason, it can also be considered the master script. Since the shot log's purpose is to assist in the creation of the final master script, which begins as a breakout script, we will examine the breakout script structure first.

The Breakout Script

Every production begins with some form of the written word. The creative director takes a written script and combines several ingredients to develop an initial vision of the finished program. He then refines that initial vision, breaks it down into discrete shots and camera moves, analyzes transitional points and defines the actual transitional effects. These may be simple cuts, dissolves and wipes, or more complex multilayered sound and picture elements.

He then steps back from the individual elements and "plays the program" in his mind to determine whether his initial choices were appropriate and to refine and embellish them. Often, he makes substitutions, rearranges elements, or creates sets of alternatives, with final decisions being left to the edit stage.

Split-format Script Notation

The split-format script breakout places audio elements on one side of the page and visual elements on the other. The split-format script may be thought of as a copy-driven format. It is convenient for several reasons. The format fits an 8½-in. x

11 in. format which makes it easy to carry. It can be created with many word processing programs (using the file containing the initial narrative script input).*

This allows the script to be easily updated. It can be expanded; the left-side visual description can be augmented by actual visuals using simple graphics programs or through more traditional cut-and-paste hard-copy methods. Its primary disadvantage is that a visual flow that carries over page breaks is more difficult for the novice to visualize than through the storyboard method. (See Figure 7.1).

The storyboard format is a more visually driven format than split-format script notation. It is, however, an excellent presentation medium. It allows the viewer to

---

*Software available from Comprehensive Video, Inc., (Northvale, NJ) that will handle the split-format script includes *Scriptmaster* for **IBM-PC** compatibles and *Cinewrite* for Apple products. *Wordperfect* can be used for split-format scripts, but is not as simple to use.

Scriptor is a reformatting package available for use with existing word-processing systems (including *Multimate, Wordstar* and *Wordperfect*). It is designed for film manuscripts and is not as useful as a split-format script. *Movie Master,* available through Comprehensive, is also designed primarily for film manuscript.

Figure 7.1: Split-Format Script with Key Elements

Wide shot from saloon doors to interior of saloon. 101A: T5-6:12:20 Includes all 5 characters. Barboy, sweeping, moves across frame with BROOM. SFX: Glasses—5:12:10 Crowd—4:10:50	**RANCHER:** You know what really gets me? (Finishes off DRINK) T3-6:02:10 Good action, line delivery soft
Two shot Rancher and Bartender. 101B: T3-2:13:50 poss.	**BARTENDER:** Shhh. Keep your voice down. (As he polishes bar top w/RAG)
T7-3:50:05 (dolly to include Newsman)	**RANCHER:** I moved to this territory 'cuz I figured I'd have more choices out here, you know, more options. But it's the same here as everywhere else. (Beat—Shakes head)

NEWSMAN: Tell me about it. I hear the same thing over and over. (Also finishes DRINK) |
| Closeup Bartender 101C: T2-6:40:02 | **BARTENDER:** What about you, Frank? You've done business with some of them other fellas. |

see from 15 to 20 images at a glance, and when all boards are laid out around a room they provide a quick sense of the entire program. However, the storyboard requires a great deal of labor to prepare and can prove uneconomical for long programs.

In many corporate programs, a combination of these tools can be effective and economical. The entire program can be broken into a split-format script notation with storyboards provided for key elements such as opens, closes, logo treatments, or special visual effects. Animation and graphics elements can also be accounted for with full visuals. This provides the best of both worlds: the split-format script serves the basic production needs of the director and his team, while the storyboard approach can enhance client presentations.

Once a format that accounts for every edit in the finished program is developed, a numbering system that will track every shot throughout the production process can be assigned. This is particularly important for any program that involves out-of-sequence shooting schedules, since a missed shot will always create major problems. For most on-camera and voice-over programs produced with the single-camera technique, the split-format script technique is quite useful in its simplicity.

The following approaches are designed to consolidate all the necessary information for the production and post-production process into one basic script. This master script can be used to account for each scene as it is recorded. It can then be used to provide all the information necessary to complete the editing of the program. A quick glance through the script as the shoot wraps will verify that no shots have been left out in the production process.

The Pagination Shot Breakout

For some projects, the director may reach a point when he knows that very few changes of the script will occur beyond that point. In such a case, he may wish to employ a breakout numbering system that is based on the pagination of the script. Page one will have shots 1-1, 1-2, 1-3, 1-4 and 1-5; page two will be composed of shots 2-1, 2-2, 2-3, 2-4, 2-5, and so on throughout the script.

Sometimes, the director finds that additional shots are required after he has composed his breakdown of shots by number. This is easily accommodated. The first shot that is added between shots 1-1 and 1-2 becomes 1-1A. If another shot is added, it becomes 1-1B, and so on.

With this simple format, it is very easy for anyone on the production to locate a given shot by name, since it corresponds to the script page and the specific place on that page. Vertical storyboard forms, which break down in this fashion, are available at most art supply stores.

The Shot Series Breakout

In many productions, the director may have to assign shot numbers before all final decisions are made and must account for the possibility of many additions throughout the balance of the production. The simple pagination breakout may prove to be cumbersome in such a production environment. Instead, he may wish to adopt a shot-series approach.

This is accomplished by dividing the script into groups of sequential shots and assigning a number series to each group. The first group will be the 100 series, which will include shots 100, 101, 102, 103, and so on. The next group will start with shot 200 and progress in the same manner through the 200 series. Again, late additions can be accommodated by creating shot 101A, 102A, 102B, and so on. If shots are added to the end of a sequence, say one that ended at 207, the new shots become 208, 209, 210, which allows for easy expansion of scenes in a linear fashion.

Each group that starts with a new hundred-series number may represent a new scene, a new type of shot series within a scene, or whatever is a convenient grouping. This technique allows for a great number of additions in the process since it is not tied to page numbers in an unfinalized script.

Copy Timing

Any script breakout should include timing of the copy so that, during the production, the director can see how long, at a minimum, a given shot should last to provide the necessary coverage for the voice-over copy or off-camera narration. This timing exercise will also help the director to note where zooms and pans fit the pacing of the program and how long their timing should be.

It will also lessen the chances that he will spend too much time creating excessive coverage on any one shot. (In commercial production, "programs" must time out to exactly 10, 15, 30 or 60 seconds. Timings within individual shots become crucially important and must be tracked in tenths of seconds. There is no place in broadcast television for a 32-second commercial!)

Leave Room for Take Information

Room should be left within the area dedicated to each shot on the master script to log the reel number and counter reading or time code for each selected take. (This will become very useful as the director consolidates his information to expedite the subsequent review and edit process.)

Before we look further at the process of constructing the script breakout—which will become the master script—we should look at what it does *not* include.

This will help define what should and should not be included in the script breakout process.

The Shot Log

The shot log should let the director know why he did not select each take before the final take, or why he took additional takes after the selected take. This should be in the form of short notes such as "walked too fast," "mispronounced a word," "light in frame," "low energy," "low batteries," etc.

Ideally, the VTR operator or an assistant should keep this log. In such a case, it is the director's responsibility to inform that person of the problems with each take. This lets the director later retrieve the decision that he made during the production. In turn, it saves him from having to review on tape all the takes that he had already decided were unusable. It allows him to compare what he considered potentially usable takes. Further, it permits him to focus on what he considered strong or weak points in those takes during the shoot.

Directorial Notes

A thorough, consistent shot log can be a major part of developing a focused, efficient and professional directorial attitude. The discipline of providing the appropriate information for the log at the time of the shoot will benefit the director by forcing him to be more decisive and more verbal concerning his perceptions. This will pay dividends greater than the simple saving of time in post-production.

It will make the talent and crew more immediately aware of the changes needed on takes, directly following each take. It will also impress talent, crew and client with the director's command of the situation, by making clear that the director has a clear vision of what he wants from each take and knows what is needed to realize that vision.

The director who wishes to analyze his own working methods and improve his skills will find it helpful to examine his shot logs. He will be able to use them to gauge his own level of focus or decisiveness.

Technical Details

The shot log should also include any changes in basic setups for camera, lighting, sound or set and wardrobe that will be helpful in maintaining continuity with future takes and will assist in the post-production phase. If color balance changes are made in the camera due to changing light conditions, this should be noted. If camera filters are added or taken away, or if gain settings are changed, this will affect the matching between different shots, and these changes should also be noted. This will decrease the amount of time spent matching takes in the edit process.

Take Numbers

Take numbers should always be noted because time code may be lost at any time, and counter numbers are inconsistent from machine to machine. The use of take numbers protects the director from losing a reference later. During the shoot, the director should resist the temptation to save time by using only audio slates. When reviewing footage, the director will find that visual slates stand out at high speeds, can be counted as they go by, and greatly assist in locating the correct take quickly.

Audio slates cannot be heard at any speed faster than twice normal without special processing equipment and are often lost in background noise. If audio slates are used alone for any reason, they should include a short burst of tone so that the tone beep will assist in locating takes and in counting takes while reviewing footage at high speeds. The shot log should include a notation of "no slate," "finger slate," "tail-slated" or "audio slate only" when these shortcuts are taken. Otherwise, the director can get very confused in the edit process and can lose an inordinate amount of time simply identifying takes. This can drain creative energies very quickly.

BREAKOUT SCRIPT DEVELOPMENT

We have covered many aspects of shot logs or camera notes. This should serve to highlight the difference between this element and the breakout script, which will develop into the master script. Knowing that a shot log will be created during the shoot, the director should develop a breakout script that will include only the information that is essential to maintaining a vision of the overall program, executing the elements of that vision while directing the shoot, and realizing that vision through the editorial process.

Initial Reading

An exercise that I have found to be most helpful begins with a continuous non-analytical reading of a finished script from start to finish. The purpose of this reading should be to get a sense of the basic audio track of the program as it will unfold in real time in front of the audience. What types of words will be heard and at about what rate? What are the basic verbal phrasings and groupings of information?

After gaining this initial impression of the sound track, put the script aside and visualize the audience, the viewing context, even the specific screen or receiver on which the program will be shown. Call back what you remember of the activities that will precede the showing of the program, drawing on all the information that was garnered through client meetings and initial research. Also, refresh your visual memory of the locations, sets, graphics and any other visual elements that are already decided as necessary to the production. Review any location and set drawings, photographs and graphics sketches. Then, put them aside so that they don't dominate your visual imagination.

Audience Identification

Next, place yourself in the position of the audience. "Play the track" of the show in your mind and let yourself quickly visualize the program. If you have trouble beginning this process, it may help to record the script in your own voice into a tape recorder, in a regular, almost flat reading. In either case, run through the program without making notes and with as little external visual stimulation as possible. The intent of the exercise is to free your imagination to conjure new and unexpected images, to let them be suggested to you in real time.

Try not to become stuck on the details of any given visual segment; instead, act the impatient viewer, who wants to see something new and interesting at regular intervals. Run through the program from start to finish without repeating sections. This often triggers unexpected ways of dealing with particular parts of the program and also will allow you to create a broad view of the program as a finished piece. That broad view will, in turn, guide you as you examine the program for structural considerations.

Structural Considerations

There is usually a structure inherent in any good script, created by the writer during the script's development. A wise director complements that inherent structure while developing substructures specific to program staging. These are the elements that can be employed to maintain a sense of pacing for the viewer, to give the program rises and falls so that the viewer is actively engaged.

Defining program structure before you begin to choose specific shots is similar to reading a map before setting out on a highway. As the driver you will have many decisions to make. Being familiar with the territory allows you to make informed decisions at each turn in the road.

The first step in the structural definition of a script consists of grouping individual segments of the program into a sequence that has a dramatic structure. In a well-crafted script, this is often a simple matter of recognizing what the writer has already accomplished.

The director who inherits a dramatically sound script may simply need to create a production design that underscores the writer's intent. In a less well-written script, the director might look for and suggest ways to enhance the program, through re-arrangement, addition of "missing" elements and subtraction of evident redundancies, that will serve to group parts of the script into segments that play off one another in a dramatic fashion.

Confrontation

There are several types of dramatic structure, but all are based on some form of building and releasing tension. The most easily recognized form involves direct confrontation between opposing forces. The basic Western film progresses from one conflict to another between "white hats" and "black hats." Each conflict is resolved in some form, but gives rise to a new showdown. Tension is built, and then released. Eventually, the climax occurs in a major confrontation.

Problem/Solution

Another dramatic form is the problem/solution structure, in which characters are confronted by problems and solutions are provided for each problem. This is a structure that often works well in a marketing presentation for a line of products—the products' benefits and features can be treated as the action unfolds, secondary to the action itself, so that the viewer is pulled into the action.

Deus ex Machina

Yet another form of dramatic structure involves an outside force which upsets the normal order of things. This allows the characters to see their everyday world in a new light, reach a new awareness, so that when order is restored, they have returned with a new insight. This is the "deus ex machina" form of exposition, often used to illustrate an abstract point. *The Wizard of Oz* is a classic example of this form of exposition.

Rhythm and Pacing

There are countless other types and combinations of types of dramatic form. The director should look for them in any script and recognize what part a given script segment is playing in the service of the form. If they seem to be missing, the director should look for ways to develop them, since they provide the key to creating rhythm and pacing.

Instructional programs can be viewed as having some form of dramatic exposition as well. They often use a tease to draw the viewer into the body of the program to get answers to questions raised. Even a corporate video memo, or "talking head" video, should be examined for its potential dramatic structure before the director chooses individual breakouts of scenes.

The purpose of examining scripts for structure is that the director can control the pacing of a program as he selects the types and lengths of shots in his breakout. When the same script copy can be covered with three shots or seven shots, the director should ask himself, "Where are we in the program; what has the viewer just experienced; and where are we headed?" and "Do I want to speed up this experience or slow it down for the viewer?"

It often helps to draw a chart of the program over a constant time line, with a rising line indicating an increase in pacing, and a dropping line indicating a decrease in pacing. Pacing can then be defined in terms of the number of separate shots per minute of script, with a shot that has pronounced camera movement or dramatic action within the frame counting for two static shots. This will give the director a guide as he begins building the breakout script shot by shot.

Creating the Breakout Script

To begin the step-by-step breakout of shots in a script, the director can review what overall pacing the material lends itself to, as well as that which his audience is accustomed to, should expect, or even can tolerate. Then, before he puts his ideas to paper, he should begin to define shot sequences in segments of the program in terms of their position in the program structure.

The director must also consider his budget constraints, as they are reflected in the amount of time available to spend per page or minute of script, and judge the possibilities for pacing accordingly. This assumes that he has developed a good sense of the realities of the shooting situation, as well as his crew and talent's capabilities.

Since the video language is primarily derived from a tradition of years of film and stage viewing, the director must be mindful of those traditions as he creates a spatial reality for the viewer. He should remember that his program will be most effective if he can create for his viewer a "suspension of disbelief," a state in which the viewer accepts the contents of the screen as a reality. To do this he should create a visual and audio flow that is consistent with traditional visual grammar so that his audience is not distracted from the program by a violation of its basic expectations of sequencing.

A "bad edit" or strange cut may not be recognized for what it is by an individual viewer, but it will serve to bring that viewer back to an awareness that there was someone behind the scenes manipulating the reality. It will break "the spell" that is essential to creating the greatest impact on the viewer.

Spatial Conventions

The typical program begins with establishing shots to set the overall sense of where the viewer has been transported to. These are usually followed by tighter shots (medium shots, then closeups), since the viewer is automatically drawn to specific parts of the frame and wants to know more about what is in them.

The overall spatial sense of a scene is often re-established at various intervals and particularly when one tight shot would otherwise be followed by one that represents a large jump in space. This maintains the viewer's spatial orientation.

Predictability

In breaking any script into a series of shots, the director should balance the need to remain somewhat predictable with a desire to surprise the viewer occasionally in order to maintain interest. Predictability should be balanced. If a program develops in a totally unpredictable fashion, the viewer will distance himself from the viewing situation as he perceives that he is being manipulated.

The surprises should come in the form of new angles, sounds and effects which are a natural part of the underlying action. Or, they should occur as dramatic "twists" resulting from changes in the program's dramatic plot. If they are well used, the technical and dramatic surprises will keep the viewer engaged throughout his viewing of the program.

Motivation

In crafting a script, a writer must choose character actions that are motivated by the background and development of each character. In bringing the script to life on the screen, the director should choose shots that are motivated by the activity that is occurring in the program.

Just as a cameraman should zoom only to underscore what the camera sees, and not simply to relieve his boredom, the director should call for new shots, moves and effects to carry the program along, not just to create more viewer interest. This delicate balance is essential to maintaining an interested, alert viewer, and the creation and maintenance of that balance is one of the most intuitive aspects of the director's craft.

Maintaining an Overview

As he works through the script selecting shots, the director should regularly step back a level and review the overall flow that he is creating. It is otherwise easy to get caught in details as he describes each individual shot, forgetting the pacing that stems from the larger view. In fact, it helps to create layers of pacing, to look at the program in terms of three or four basic movements, and then to subdivide these into four or five segments each.

Scene Description

Some directors are excellent artists and are blessed with the ability to quickly translate their visions of scenes into easily understood drawings. (Frederico Fellini, for example, was a cartoonist before his career as a film director.) Others are not so talented, or simply lack the time to draw all the individual frames or the budget to hire someone to draw them. Verbal descriptions can be quite adequate when there are no sets to construct and the visual elements are easily understood by director, client and crew. Verbal descriptions are far easier to update than drawings, as well.

However, when budget and time are available, the director derives great benefits from having each shot available as a fully drawn picture. It allows him one more level of pre-visualization and is an excellent medium for resolving questions with clients long before the production begins. In some instances, reproductions of the drawings can be stacked to provide a fast review of shots in sequence by flipping the edge of the stack—the simplest form of animation, but highly effective in that it allows the director to experience the visual flow of the program in a compressed time frame. This technique also makes bad edit decisions evident before the footage is shot.

Scene Element Notes

The director should make notes on his script regarding elements that will be involved in upcoming scenes, shot sequences and setups. These would include props that are needed, particular styling and lighting considerations, changes in camera mounting, talent required, specific audio needs, etc. Copies of the script and notes can then be distributed to the persons in charge of props, lighting and sound, and they can plan and even prepare future setups in the course of the day.

This allows the crew to prepare each setup much faster since conversation can focus on the details of execution rather than on description. It is also helpful because it forces the director to visualize the particular actions needed to set up each shot, which, in turn, allows him to assign the proper amount of time to each setup. It may even lead him to modify a shot or setup as he weighs the time involved against its relative importance in the script.

SHOT GROUPING

After creating the script breakout (which will account for every new change of shot in a program), the director should next begin to organize the production itself by creating a shot grouping that organizes the shots in terms of "setups." This process sometimes works best when it is done in two steps.

Setup Groups

In the first step, the director places shots into various setup groups as he works his way through the script. Each setup can represent a move to a new location: from one room in an interior location to another room within that interior location. It may represent a change from an interior location to an exterior location. Or it may represent the change from one exterior location to another exterior location.

The setup may also represent a major change in lighting, camera mounting, or any other change which will cause a major break in the final shooting schedule. The list he creates will thus be broken down into groups of shots that will be done at one time, but still maintains the sequencing in the finished script.

Shot Sequencing

In the next step, the director really focuses in on the logistics of the production process. He examines each shot within the setup grouping to see where it might be placed for maximum production efficiency. He is trying to generate as much creative time for himself as possible by minimizing the number of times equipment movement is required.

For instance, if an actor is to be shot in a scene while working at a keyboard on a computer terminal, the director might choose the following series of shots:

1. a wide shot of the actor and terminal, looking over the terminal at the actor
2. a close–up on the screen of the terminal
3. a close–up on the keyboard with the actor's fingers
4. a close–up on the actor's face
5. another close–up on the terminal screen
6. a close–up on the actor's face
7. another close–up on the keyboard
8. finally, an over-the-shoulder shot of the actor that looks into the screen of the terminal.

One approach to sequencing these shots may be to do shots 8 and 1 first, since they involve the most elaborate lighting and possibly the most complex moves, then shots 4 and 6, since they are from the same angle as shot 1, but with tighter framing. Then, shots 3 and 7 since they still require the talent, but only his fingers. Finally, shots 2 and 5, which may require relighting but will not require the actor, who can be released from the set (thus saving talent fees). Figure 7.2 is a sample shot list.

Figure 7.2: Sample Shot List

Project: VOICE COMPANY
Computer Telephone System

Secretary's Office

16-F MS Sect'y puts on headset and begins typing
17-B MS Sect'y typing., answers phone on 2nd ring
18-A MWS with Ubangi for empty desk dolly
16-D MCU Sect'y puts on headset
17-D CU Sect'y answers phone call
18-D MS Sect'y finishes typing and sets up for shared file, then print.
21-C MS over left shoulder OTS Sect'y gets Don's call & responds.
16-E Footswitch CU
17-A ECU screen with typing as phone rings
18-C ECU screen—typing continues after phone interrupt (looser)
17-C ECU phone window with "IN USE" for incoming call
18-E ECU screen with important file notation
18-F MS printer typing Don's letter

SEQUENCE:

All screen close-ups, then cross axis (21-C), footswitch, then all MS, MCU, CU of Sect'y
    end with 18-A to lead to
18-G Narrator enters

Customer Service Area

1-A Open with Narrator
22-A Closes with Narrator
22-B CU screen for close
2-B Open segment 2
3-A CU on Computer Telephone System for Key Points
3-B New angle for intro to demo
Voice-Overs
13-B Drafting area open with "annotation"
15-A MCU in drafting area for dictation throw

Additional screen close-ups if required

Sandy's Office

8-E OTS Sandy as she works on spreadsheet & gets call
9-B MS Sandy as she gets Don's call
10-A (9-B cont'd.) MS as Sandy finishes call, hangs up and changes screen
11-A OTS WS continuation of 10-A
11-E MS Sandy exits graphics and sets up voice
12-B CU Sandy annotating report
14-C OTS Sandy sends message and returns to spreadsheet

Sandy's Screen

11-B Spreadsheet changes from original to Reardon—MCU
11-C ECU spreadsheet (Reardon) numbers change, then blank
12-A MCU VoiceEditor window appears over Reardon document
14-A MCU Reardon document with voice Icon visible, screen blanks
14-B ECU address screen—Don's name is filled in

Sequencing within the Production Day

The above approach dealt with the major lighting and critical performance first, then with easier shots, and finally with shots that did not require a key element. This approach would be particularly appropriate if the setup were late in the day, and the crew was already "up to speed" to tackle the most complex shots first, then the easier shots, and finally, the simple shots not requiring the actor.

If this were the first setup in a shooting day, the sequence might be reversed. The talent could be in makeup and wardrobe or reviewing lines while the screen shots (2 and 5) were done. This would allow the crew more time to think about the lighting for the larger-scale shots (1 and 8), which would come later, require more instruments and more "tweaking." Crew and director would have a start–up period in which the miscommunication and mistiming that occurs naturally in the early stage of a shoot would not create a major problem.

Other concerns aside, one basic rule of thumb in the industry is to begin with tight shots and work out to wide shots. This allows a good continuous flow of production action, since the tight shots are usually done quickly and equipment is added in small steps as the shots become larger in scale. Of course, the director must present the "whole picture" to the talent and crew at the outset so that a consistency within a scene is maintained. Another approach, is to begin with the widest shots to establish that whole picture, and then to do the tighter shots within the overall picture.

Location Considerations

Other considerations for the sequencing of setups throughout the day might begin with the most basic ones. Where do I want the sun to be during this shot, to suggest time of day or to create a mood, or where does it have to be to even make this shot doable? Are there concerns relating to location activity that dictate the time when the shot is recorded? The most obvious situation may be the one in which commuter activity is part of the scene, or when particular company personnel must be available.

Background sound may ebb and flow during the day so that sync sound should be done at particular times. If the setup is near an airport, flight patterns should be studied to minimize the number of retakes and reduce the stress on the talent. If traffic flows nearby, rush hour might best be avoided for the same reason.

Sometimes subtle factors come into play. If a client's computer system is part of the program, the director might want to film screen activity at a time when it is not being used so that the system's response time appears faster.

The director may know that a particular actor performs best early in the day. Or, he may know that the actor is in a stage production at night and will appear tired if scheduled for an early call. He can adjust his production schedule to elicit the best performance from his talent.

Personal Preparedness

Each director may have his own reasons for choosing a particular sequence of shots within a setup, and may have particular reasons for arranging setups in a specific order as well. No matter what the reasons for those choices, though, the director gains from the process, since it forces him to rehearse the production day itself. This will prove invaluable.

Operating with a fully detailed script breakout and a finished shot list, he will be able to answer quickly the question "What's next?" at any stage in the production day. He will also be better prepared to anticipate production problems, since he will have already "been there" in his own mind.

Weighing Key Elements

Ultimately, the director must act as a resource allocator. He must choose to rank some shots as more important than others. Often, opening sequences set a style or tone for a program and merit more complex treatment than other elements of a program.

Sometimes a product is introduced or highlighted at a given point in a program and the sequence involved justifies a more spectacular treatment. The same may hold true for the closing portion of a program, since the final image sequence is often the one the viewer remembers most strongly. The director may choose a production sequence that allows him more creative time with those key elements.

Providing Adequate Coverage

The director also must ensure that he has enough "coverage." This means that he must approach the editing session with sufficient shots to cover the copy—that there are no holes in his production plan and shot selections. This involves not only timing the script to ensure that no shot runs short, but making sure that each shot is pre-visualized adequately to ensure that there are no uneditable sequences.

He must also allow for occasional backup shots that will substitute for shots that are planned, but which his experience tells him might be difficult to execute successfully. These may be shots that involve elements on the location that are not completely controllable. Or they may be shots which hinge upon the performance of actors who are not well known to him. They might also be those shots that require extremely precise timing among crew members.

Daily Rundowns as Checklists

The finished shot list contains only groupings of shots by name and the order in which they are to be shot. From this list the director can create a one- or two-page list that can be attached to the front of his breakout script. This list will quickly indicate, for each day, which shots will be done in each setup, and in what order.

This daily rundown can also serve as a primitive checklist. As each shot is completed, the director can draw a line through its number on the rundown sheet. This will ensure that setups are not struck before all shots planned for that setup are completed. Another simple but effective way of tracking shots is to draw a diagonal line through the copy side of the script, so that a quick scan reveals any unmarked passages.

I have found that a particularly effective way to track what is done and not done is to combine the two methods to provide a form of cross-checking. As you complete each shot, draw a line through the text on the script first, then mark off that shot number on the shot list attached to the top of your script. When all shots on the shot list are ticked off in sequence, double–check the script itself before striking the setup.

Avoiding Embarrassment and Wasted Effort

The use of the above methods helps to avoid some of the worst possible moments for the director. If, for example, the director has told the crew to remove the lights and strike the setup, and he has released the talent, who then break character, it is distressing if he discovers that he must say "Oops, I was wrong, set everything back up the way it was, we have just one more shot to do."

This makes extra work, tires the crew and talent, and makes everyone suspicious of the director's command of the situation from that point forward. Only the most self-confident director can bounce back from one of these moments to focus on the direction of the next scene. The less experienced director should adopt a very methodical mode of shot checkoff, even if he thinks that the crew seems impatient with the process. The director should not let his ego or lack of self-confidence get in the way of his saying, "Stand by everybody, I think we're done here, but I need a moment to cross-check my notes."

A complete shot log or set of camera notes kept by an engineer, VTR operator or assistant is very helpful in this regard. If the director is confused by his own notes, he can verify what has been shot and what hasn't been shot by an examination of the sequential log. In the event that the question isn't answered through the log, he can review the footage which has been shot. However, this should be a last resort, since the process is time-consuming and reduces production momentum.

Creating the Production Schedule

Once a script breakout and shot list are created, the director has the basic elements in hand to construct the final production schedule (see Figure 7.3). To do this, he must first assign a specific amount of time to each shot or set of shots. He should proceed through his shot list in the order that the show will be produced, so that he can again visualize the entire production sequence.

At specific points he should again review how long each block of shots will take and relate it to the hours of the production day. Typically, this is done in a multi-

Figure 7.3: Sample Production Schedule History Video

3-16	Pre-production script review and objectives assessment
3-17	Initiate visual research with client Script review
3-19 & 3-20	Music and narration search
3-23	Final script presentation and visual selection
3-24	Visual preparation for motion control
3-25	Narration session Motion control transfer session
3-26	Window dubs struck
3-27 to 3-30	Off-line edit
3-31	Off-line edit presentation
4-1	Off-line re-edit as required
4-2	Graphics preparation for animation
4-3	Animation session
4-6	On-line edit session

Note: This was used with a simple historical production that used existing footage.

step process, since the first time through the script usually creates a production schedule far longer than the budget allows. No director likes to work faster than he must, but the realities of the production world usually dictate that he must choose where to allocate a limited amount of time.

After blocking out the time desired for each shot, and finding that he has created a production day several hours longer than allowable, the director goes through the script again. In this step, he takes a small amount of time away from each shot series, until he creates a nominal production schedule that allows him the best compromise. In this process, he must again weigh the relative importance of shots as he reduces the time allotted to each of them.

The Final Production "Bible"

The production schedule should be completed as a final production "Bible" which includes each step in the process of the production day and any helpful support information (Figure 7.4). A typical format would include:

1. Arrival time of crew on set or location.
2. Additional arrival times for crew persons not required immediately (make-up artist, stylist, prop person).
3. Talent arrival time.
4. First setup time highlighted so that the crew knows when they have to be "up to speed."
5. Requested client arrival time: this should be no sooner than necessary so that crew and director can speak frankly and won't have to explain what they are doing any sooner than necessary. (The initial stages of a production are confusing to the outsider and inevitably generate a great many unnecessary questions that are self-explained once the production gets rolling.)
6. Meal breaks at reasonable points.
7. Times when experts, consultants and special props are required.
8. Finish time for talent; talent "release time."
9. Strike time of final setup.
10. Time out of the location or studio.

FINAL NOTES

The production schedule should be prepared in advance of the shoot and the director should have a second person familiar with the production review it for possible oversights. Once confident of a realistic schedule, the director can run through the production day one last time—a final rehearsal to pinpoint the opportunities for specific creative refinements within the defined limits.

Figure 7.4: Sample Production Bible

POSITION: PERSONNEL

Line Producer/AD: Tom Kennedy
2nd Assistant Director: Jane Doe
Script/Continuity: Jane Smith
Director of Photography: John Jones
Gaffer: Gary Schwartz
Best Boy: Tim McCoy
Key Grip: Jeff Goldberg
2nd Grip: Chris Smith
Sound Recordist: Greg Gonzalez
Home Economist: Sandra Doe
Asst. Home Economist: Kathy Jones
Makeup Artist: Annie Smith
Stylist: Gail Jones
Production Assistants: Jan Smith
                                    Jim Jones
Video Technician: Milt McCoy
Welfare Worker: Renie LaFrance
Sets/Props: Technique: Mary Doe
                                    Jim Smith
Catering: Abby Doe

KEY PHONE NUMBERS AND ADDRESSES

Advertising Agency or Client
999 Financial/Creative St.
Their City and Zip
Their Phone Number

Our Production Company
111 Efficient Ave.
Fun City and Zip
Office Phone Number
Producer Home Phone Number

Location:

(Vacant storefront)
207 Corte Madera Ave.
Corte Madera, CA 94925

One block from the point at which Magnolia Ave. becomes Corte Madera Ave.

Diagonal on-street parking available opposite location storefront.

From San Francisco: 101 North to Corte Madera exit, west (left) to Magnolia, left on Magnolia to Corte Madera Ave.

Figure 7.4: Sample Production Bible (Cont.)

PRODUCTION SCHEDULE—TUESDAY, 10/7

7:00 AM—Crew Call at Location

7:45 AM—Makeup/Wardrobe Arrival

8:00 AM—First Talent Call

8:45 AM—FR 4, 5 Switchboard (Marie Doe—8:00)

9:30 AM—FR 3, 4, 6 Aerobics (Wendy Smith—8:30)
Brick Wall Flat (A)—Counter (Product Pickup)

10:15 AM—FR 7, 8 Aerobics (Mary Jones—9:00)

10:45 AM—FR 9 Switchboard (Susan Evans 9:30)
Kitchen Print Flat (B)—Product—Loaded Fork

11:30 AM—FR 9, 10 Aerobics (Sarah Doe—10:00)

12:00 PM—FR 8 Switchboard (Laurie Smith—10:30)
Grasscloth Wall Flat (C)—Counter Pour—Phone—Naked Fork

1:00 PM—Lunch Break

2:00 PM—FR 10 Switchboard (John Doe/Jack Smith 11:30)
Floral Flat (D)—Counter w/Covering—Foreground Phone Product & Hero Salad
Foreground

2:30 PM—FR 11 Aerobics (Class Extras 1:00)
Foreground Product—Salad—On Unseen Counter

3:30 PM—FR 1, 2 Aerobics (Town Extras 1:30)
Ext. Aerobics Sign—Principals and Class Extras Inside

RELEASE PRINCIPAL TALENT

4:00 PM—ALTERNATE FR 1 Aerobics (Acute angle LS)

4:30 PM—FR 1 Times
Times Sign (Extras Only)

5:00 Wrap is Called—EXTRAS RELEASED

5:30 PM—Crew is Dismissed

Figure 7.4: Sample Production Bible (Cont.)

PRODUCTION SCHEDULE—WEDNESDAY, 10/8

7:00 AM—Crew Call on Location

Talent Calls: 8:00 Karen Doe (Rprtr 2)/ 8:15 Ellen Smith (Bagger)/ 8:30 Ed Doe/
9:00 Susan Evans (Rprtr 1)/ 9:30 Scott Smith/ 11:00 Eric Jones (Newsboy)/
1:30 Ann Smith

9:00 AM—FR 11, 12 Times
Desktop—Hero Large Bowl—Newspaper Drop

9:30 AM—FR 4-5 Times
CU Hero Product pour with talent in BG

10:00 AM—FR 2-3 Times
Master Open with editor—Bag Woman Entry SYNC

10:45 AM—FR 6-7 Times
Salad—Product Hero—Loaded Fork (Reporter takes bite) SYNC

11:15 AM—FR 8 Times
2nd Reporter—Product

12:15 AM—FR 10 Times (Ext./Int.)
Newsboy—Newspapers SYNC

RELEASE SERIES 1 TALENT

1:00 PM—Lunch Break

2:00 PM—FR 2 Switchboard
EXT. Mullion Window, Sign, Switchboard, Headset

2:30 PM—FR 6 Switchboard
Interior Operator in closer shot than FR2

3:00 PM—FR 11 Switchboard SYNC
Hero Salad—Product—Loaded Fork

RELEASE SERIES 2 TALENT

3:30 PM—FR 12 Switchboard
Static Hero Product—Background Salad—could become hero salad dependent on focal
length chosen

4:30 PM—Wrap Called

5:30 PM—Crew Dismissed

Figure 7.4: Sample Production Bible (Cont.)

PROPS & SETS REQUIRED

Flats:
    A—Brick Wall 4 x 4
    B—Kitchen Print 4 x 4
    C—Grasscloth 4 x 8
    D—Floral Print 4 x 8

Counters:
    Oak Formica
    Print "Tablecloth" Wallpaper covering

Signs:
    Times 3-ft. x 4-ft.
    Telephone 2-1/2 ft. x 4-ft.
    Aerobics 2 cloth signs:
        Done on Site

Aerobics:

Dance Bar for Aerobics Class

Switchboard Operator:

Console (working)
    Headset
    Chair

4 Modern Phones (1 a Rotary Desk Type)

Times:

Bulletin Board	Newspapers (Many)	3 Desks
3 Chairs	Typewriters (Manual)	
	(1 an Underwood Upright Style)	

Wire In/Out Baskets

This process will pay enormous dividends in terms of professional self-confidence and focused directorial performance. Crews and actors respect the director who has done his homework since it allows them to work up to their full creative potential.

A further dividend comes from client perception. Clients are usually impressed with the director who has shown respect for their budget and has delivered a polished production plan. They arrive at the set with increased confidence in the director and his team. This combined set of attitudes can create a very positive atmosphere within which the director can work well.

The Producer/Director Tension

The process described in this chapter reflects the two roles that the director of a small crew must play. On the first pass through the script, he acts as a traditional director, simply creating the best possible program from the materials at hand. On the second pass, he is playing the role of a producer, or production manager, who must bring the program in on time and on budget.

This duality is a constant source of tension for the producer/director. However, the best way for him to maximize his role as a creative director during the shooting process is to act the stern producer prior to the production, so that he constructs a thoroughly realistic and fiscally responsible schedule. Then, within the time limits of that schedule, he can focus completely on the job at hand. He can concentrate fully on his directorial task during the production day.

Without a well-planned schedule, the director will find himself thinking too much about the timing logistics of a shoot at the moments when he should be looking closely and critically at the performance of the talent, the timing of a camera move or the aesthetics of the composition of a particular shot. Most directors who have been given the opportunity to work with a separate producer, production manager, or assistant director develop a clear sense of this duality. They fully appreciate the separate role of the "producer-type"—the person who keeps them on schedule by telling them how much time they have left to do a given shot.

In the absence of this resource, the director must learn to switch gears just prior to a production. As his own producer, he can create a final production schedule that he can live with as a director. This at least lifts many of the moment-to-moment logistic concerns from his mind when he again switches back to his director role.

8 Directing Non-Professional Talent

There are many production situations in which the director is asked to direct an on-camera performance by someone other than a professional actor or spokesperson. A message regarding the future of a company may be delivered best by the actual chief executive officer. A technical program might gain great credibility with its audience when specific facts or findings are presented by the person responsible for their discovery. A medical presentation may gain greater acceptance by its viewers when the spokesperson is a medical professional. Or, budget parameters may simply prohibit the use of a professional actor.

In any of these instances, a director may be asked to undertake one of the most challenging, and potentially unrewarding, tasks—directing the performance of someone who does not normally perform for the camera. This can lead to an awkward situation due to a variety of reasons over which the director may have no control. Or it can be very rewarding for the director who enjoys helping people to perform at their best.

DIRECTION VERSUS CRITICISM

The most common cause of problems in the performance of nonprofessional talent has to do with their perception of the director's suggestions and critiques. The talent may have no previous experience that would allow them to accept direction as something other than personal criticism. In such an instance, they see the director as an "outsider" to their organization with few credentials that would allow them to trust the director as someone who has their own best interests at heart.

As a result, they may treat every suggestion made to them with distrust. This will stand irrevocably in the way of a director who is trying to assist them to give a good performance. Creating trust is essential to building a working relationship between director and talent.

Building Trust with the Talent

For a director-talent relationship to work well, the talent must trust the director as the person to whom they will listen intently and whose direction they will follow, almost without thinking. This relationship is important because, in the act of performing, the talent must not be too self-critical or too self-aware. The nonprofessional talent must act naturally to be believed. The performer who is too "outside himself" (the natural result of constant self-criticism), projects an unnatural presence which undercuts his credibility with the viewing audience.

The director should act as the ultimate judge for the talent. This allows him to give immediate feedback, which can be incorporated by the actor without having to "break character" and become his own critic. The key to working well with nonprofessional performers lies in building this trust relationship *quickly* so that the talent get the maximum benefit from the director's expertise, before the talent become too tired to give a naturally energetic performances.

The Pre-Production Meeting

There are several techniques available to the director that will enhance his chances of creating a trust relationship quickly. The most obvious of these is to approach each actor on a person-to-person level outside the production environment. By doing this, the talent can assess the director as simply another human being. Most important, the nonprofessional talent should perceive the director as a person who understands, agrees with and shares the talent's goals for performing. (This factor does not figure so centrally in a director's relationship with the professional actor, for reasons we will discuss in the next chapter.)

To enhance the rapid development of a fruitful working relationship, the director should try to arrange an early pre–production meeting. Prior to an on-camera performance by a chief executive, the director can schedule a meeting simply to discuss the production.

The Importance of Working Environments

Such meetings allow the director to address the talent in their own environments, where they are accustomed to being in control of the course of events. In such settings, the director will be able to introduce himself as someone who is interested in assisting the executive, or technical expert, in meeting his own objectives—in fulfilling the goals that impel him to appear on camera in the first place. With this as a starting point, the executive or technical expert is most likely to begin to trust the director. It helps to project little "directorial attitude" on such occasions and to show as much interest as possible in the nonprofessional performer and his concerns.

Assessing the Talent

This initial meeting also allows the director to assess the experience level of the nonprofessional performer and discover the talent's strong and weak points. If the talent has a speech pattern that accentuates sibilance, a script might be modified to reduce the number of times those sounds occur. If the talent has habitual patterns of movement that tend to distract, the director can begin to strategize staging, camera movement and talent movement to draw attention away from such distractions.

One of the first questions to the talent should be, "Have you appeared on camera before?" The director should never assume that the talent does or does not have a particular experience level. Both over-assuming and under-assuming can create problems.

If a nonprofessional has been on camera before and has had a terrible experience, it will help greatly to know what aspects to avoid. Perhaps the lighting accentuated a bald spot, bags under the eyes or some other feature—items which can be easily dealt with. The talent is unlikely to discuss these concerns in front of any-one in the production situation. By addressing these concerns in a private, personal fashion, and prior to the production, the director will strengthen the working relationship.

This personal querying process will also help to avoid another common problem—the hidden concern that causes a talent to be noncooperative in the production itself. Many times, when a production seems to be proceeding smoothly, the nonprofessional may become difficult to work with, distracted, and unwilling to take direction. If no prior contact has been made with the talent, this may often be his way of expressing objections to production decisions or elements which he does not feel comfortable with confronting directly in a public situation.

PRODUCTION DETAILS

A second, key goal of the pre-production meeting is to define the specific requirements for the performance which the talent is required to undertake. This should include every detail of the production.

Using Notes, Cue Cards and Prompters

If the talent insists on using written notes, the director should know this and be prepared to deal with the possibility of a seemingly "read" performance. This will also create a lighting concern which can be dealt with at this time. Since white bond paper with notes will be a distraction, the director can request that the talent have his notes finalized prior to the taping so that they can be copied on a paper stock more suitable for television. The hidden benefit of this is that the talent is forced to prepare notes early enough for the director to review them.

If cue cards are to be used, the director can discuss the logistics of their use, particularly if the talent is not used to the technique, at this first meeting. Again, this allows the director more time for preparation, since he can ask the talent to do a thorough run-through with the final cue cards prior to the actual production. This also gives the director more time to work with the talent before the taping and invariably results in a more successful performance.

If a teleprompter is planned, again the director may request that the talent's script be finalized or approved in time to prepare it for the prompter. The talent may be asked to schedule a prompter rehearsal. At this meeting, the director can ask the talent about eyesight constraints, since this often affects the successful use of cue cards and teleprompters. If the talent requires glasses to use the prompter or cue cards, the director can plan accordingly with talent and crew.

Eyeglasses

Discuss the use of eyeglasses before you get to the set. Glasses are troublesome to deal with in terms of lighting and often a crew will ask if the talent can appear without glasses. It is difficult to decide at the last minute and may directly affect the talent's performance. He may end up not wearing glasses in deference to the crew, while being unsure of himself without them, or wearing them while worrying about his appearance. Neither option would serve to support his appearance as a calm, confident spokesperson.

Props

Other items that should be discussed at this initial meeting include all prop considerations. Too often, a novice on-camera talent appears at the studio with a number of props that work perfectly well in his speeches but are inappropriate, dated or simply unworkable for television. The experienced director should be aware that to remove such props at the last minute will greatly reduce the talent's self-confidence, sometimes causing inability to perform. The talent may not know how to regroup and reorganize his spoken material on short notice.

To avoid this, the director should deal with all prop concerns at the outset of the production planning process, so that he can reach an agreement with the talent for the incorporation of elements of visual support.

Charts and graphs can be treated as full-screen visuals that the talent will not handle but can refer to. Often the talent needs reassurance, since he is accustomed to handling the props, or needs them as a memory jog. You may need extra rehearsal time or have to suggest techniques to help the talent compensate for the loss of his usual props.

Wardrobe

At the initial pre-production meeting, the director should address the question of wardrobe selection. He should assess the nonprofessional talent's personal tastes. Then he can tactfully suggest guidelines for on-camera wardrobe selection.

These suggestions should, of course, include warnings regarding stripes, checks and plaids. In addition, a fair-haired, light-complexioned talent should be advised to avoid too dark a wardrobe, since the harsh contrast between his own coloring and his wardrobe will prove subliminally distracting. Conversely, a very dark-haired, deep-complexioned talent should be asked to avoid white shirts and light suit jackets.

By including the on-camera performer in this aspect of the pre-production process, the director can give the performer a greater sense of self-confidence. Specifically, he can set up a "win-win" situation in regard to wardrobe. To understand this more fully, let's first picture the losing situation.

Imagine that the executive performer is not told of any wardrobe concerns prior to a taping engagement. (This happens all too frequently.) He appears at the studio or location in a distracting patterned jacket or ultra-white shirt. He hasn't brought any optional choices. The technical director or lighting director immediately points out the problem and asks for additional time to adjust lighting or camera setup to compensate. At this point, before any on-camera rehearsal, the performer has been sent a message that he has already made a mistake. A losing pattern has been initiated.

On the other hand, imagine that the same person appears at a studio in the wardrobe that was selected with the guidance of the director in a pre-production meeting. The crew immediately indicates to him that his wardrobe selection is perfect for television. The performer appears and feels knowledgeable from the outset of the production situation. His self-confidence is bolstered and he is able to focus on the next step in the process. Further, his trust of the director is reinforced. By following the director's advice, he has been made to "look good."

This may seem a relatively simplistic example, but the principle is a proven one. A concerted effort by the director to make the nonprofessional talent look good in the early steps of a production will pay off handsomely in terms of a better on-camera performance.

Makeup

Some nonprofessional performers who are successful in their areas of expertise have become so without spending much time or energy in developing their personal images. Others may have developed an awareness of their image and have taken steps to improve it. The current trend is toward greater attention to one's personal appearance as a part of advancing one's career.

Gender-generated Attitudes

Generally, women have developed a more refined sense of their image than men, and they are accustomed to utilizing makeup to enhance their appearance. Consequently, the use of makeup poses different concerns depending on whether the director is dealing with female performers or male performers.

If the director has the budget for a makeup artist, he may find that women are most receptive to having one provided for them. The director then simply asks them to appear at the taping site with as little makeup as they feel comfortable wearing in public. If the budget does not allow for a makeup artist, the director may suggest slight modifications to the person's normal makeup. The director should be aware that it can be a sensitive area, since the woman may have developed a personal "look" and may be resistant to any change. Emphasize that the change is for the camera.

Less Is Usually More

Usually, the director will find problems with an overly made-up look rather than a too-sparing use of makeup. The camera will accentuate this and undercut a woman's credibility with the audience, since the look itself will become distracting. Most often, the best approach is to suggest a light hand, with the assurance that adjustments can be made after a camera check. Once the performer is lit and in position, a small amount of tape can be recorded and she can be the judge of her appearance on camera.

When using this technique, it is important that the director conduct the on-camera test with the performer in action. Otherwise, the actor's review may easily become too self-critical, treating the medium as if it were the same as still photography. In fact, if the performer seems to become overly critical or unsure, the director can use the fact that television is a motion medium and that the performer is always his own worst critic to provide some reassurance.

Male Makeup Concerns

Often, a basic makeup will greatly enhance a male performer's appearance. Many people lose much of their personal power when it is translated to the screen if they have distracting features which will be accentuated by the camera in close–up. (The classic example of Nixon's appearance in his unsuccessful campaign debates immediately comes to mind.) The director should be attentive to such possibilities and should be prepared to persuade the resistant male performer to take advantage of the benefits of makeup.

The director should treat the use of makeup as an accepted and standard practice. If he does encounter resistance, the director may try a less direct approach.

One approach that often works is to ask the performer if he would like to look as if he had just returned from Hawaii. This implies that makeup will simply be an enhancement and will not appear unnatural. Recounting examples of persons who have used makeup in past productions can prove helpful as well.

When the budget does not allow for the inclusion of a makeup artist the director may wish to use the most basic of makeup approaches with a male performer. The director should ensure that a basic makeup kit is available at the time of performance and that someone on the crew knows how to use it. Just in case, the director may wish to become versed in the basics and bring his own kit to the production.

Basic Makeup Needs

A basic kit will consist of a moisturizer (to ensure that other products are applied evenly), translucent powder, camel hair brush and protective bib. Optional products would include a bronzing gel and a blushing gel. Any good-quality cosmetic products can be used.

When there is no makeup artist, the personal contact that is involved in doing the performer's makeup gives the director a good sense of the performer's state of mind and furthers the sense of trust between performer and director. When a makeup artist is within the production's budget, the selection of the appropriate person can greatly assist the talent's performance.

The Makeup Artist's Role

A good makeup artist will be sensitive to any concerns that performers may have about their appearance and will help them adjust to any new image. If the performer looks good but feels uncomfortable, the purpose of using makeup is defeated.

A makeup artist who is outgoing and shows genuine interest in the performer creates a relaxed atmosphere for the performer. This can help immensely since the time when they are working with the performer is the time just prior to an on-camera appearance. The makeup artist can entertain, amuse and distract the performer who might otherwise become tense. This is particularly important in a production that involves complicated setups, since the talent will be waiting for extended periods of time between setups.

Matching the Set to the Person

Just as the nonprofessional may hesitate to use makeup, he may have similar concerns about the environment in which he is seen on camera. At the first meeting with the performer in his office I can get clues about his environment. In turn, I can judge what setting will reinforce his sense of comfort by matching the set design and colors to the general feel of his office. I can also discuss set elements in general terms so that the performer isn't surprised and distracted by an unexpected setting on the day of taping.

Psychological Concerns

Throughout this process, the director should remain very sensitive to his own role in relation to the nonprofessional. He should appear confident and knowledgeable while never denigrating the talent's inexperience and lack of knowledge, so that the talent continues to trust his judgment. Whenever the talent puts forth an unworkable idea, the director should first try to understand the talent's motivation and then tactfully suggest an alternative method for effecting what the talent wishes to accomplish. This way the talent will continue to perceive the relationship as a team concept rather than an adversary relationship.

Further, the director should at this point develop a "win-win" strategy for the performer. He should give the performer the information that he needs, when he needs it, to boost his confidence as the production approaches.

Much of the success of such a strategy depends on the director's ability to put his own ego aside and always deal with the talent in a supportive manner. He should also set an example for all crew members who become involved in the process so that they do not flaunt their expertise or belittle the talent's lack of expertise.

THE PERFORMER'S SUPPORT SYSTEMS

At the pre-production meeting, the director can note who the persons around a nonprofessional performer are and how they interact. This will give him clues which will prove helpful in shaping his own directorial style to best fit the performer's expectations. Often, in dealing with corporate executives, the director encounters a particular personal style or corporate culture. If an informal style and attitude prevails, the director can guess that this will be a good working style for the production; but if a very formal style prevails, he may wish to adopt a more formal working style and perhaps suggest the same to his crew.

The Office Structure

If the production is slated to take place on location at an executive's own offices, the director should take the opportunity to meet the executive secretary or any other key "gatekeepers." Many times the executive doesn't schedule his own time, or even control his own office environment, and the director will wish to enlist the aid of the executive secretary to help schedule production elements.

Just as important, the director may find that the executive often changes schedules or has them changed, and the videotape schedule is considered to be a much lower priority than a sudden board of directors meeting. In such cases, having an ally within the office of the executive can prove very helpful in minimizing unpleasant (and possibly expensive) surprises.

On Location at the Office

Having allies within an executive's office can also be very important when dealing with location logistics. The director who uses an executive's office as a set will want to set up and light in advance of the taping—preferably while the executive works elsewhere. This is important so that the executive is not distracted and annoyed by the setup process.

Bringing in lights, cables and equipment, moving furniture and adjusting plants and drapes are things that the director and crew are quite accustomed to. Executives, however, are typically accustomed to controlling their own environments and may react very unpredictably.

Productions work more smoothly if the executive is allowed to work in another office until needed. When the executive returns to his office, the office is now a "set." The director is in control and no power struggle need occur, so long as the executive is reassured that order will be restored.

Creating Corporate Allies

Most executive secretaries will be more than cooperative when the situation is explained to them. In order to enlist their aid, the director should explain the overall production and should be specific about the nature of the process.

Since most people outside the production industry have little or no idea how much equipment is involved, I have gone so far as to show them production stills which include lights, C-stands, dollies and video equipment in action. This allows them to be "insiders" who are part of the production. It also prepares them for the scope of the intervention and to remain helpful in the face of chaos. Again, maintaining their trust is important, since their first loyalty is to the company and its executive rather than to a director who intrudes on their lives for a day and then disappears.

The assistance of office staff can prove to be invaluable as a production unfolds, since the crew may need help locating power panels, finding props on the premises, and finding janitors or electricians during the shoot itself. The few minutes spent in the early phase of production to "include the support persons in" will prove invaluable when time pressure becomes critical in the course of the shoot. The assistance of persons who know how to get things done within a company's infrastructure can save a great deal of time, and sometimes save the shoot itself.

Onlookers

The director also needs to address the question of which persons in a company will be included during the taping process. Often executives who are unsure of themselves will wish to have an audience of trusted onlookers. Sometimes, these persons will invite themselves. I have at one time had 10 different persons present in a control room while the company president taped a simple video memo. These ranged from secretaries to vice presidents and members of the legal department. Needless to say, the presence of unnecessary onlookers can create difficult situations for director, crew and executive as well.

The director should recognize that some onlookers may be genuinely concerned about providing moral support for "the boss," but many may have other motives. I have found that the executive should be approached directly at the outset of the production and be asked whom he or she would like present during the taping.

In this way, the director can work with the executive to control the situation and use the executive's wishes as the reason for excluding unnecessary persons. Without executive sanction, the director may have to put up with the situation or risk being perceived as a high-handed character who is controlling access to the executive.

The fewer onlookers there are present during any nonprofessional's performance, the greater the chance the director will have to shape the performance for the better. Although the nonprofessional may initially feel that the support of an audience will boost his confidence, in fact, as a director tells him what to do to improve, he will be caught in an awkward situation. Normally, an outsider is not "the expert," he is. Normally, he doesn't take direction, he gives it. Normally, he is in control of the situation, not controlled by it.

In this case he will start to consider how he is being seen by fellow workers or employees. This will tend to get in the way of his acceptance of the director's suggestions. He will act less instinctively and react less openly to suggestion. The net effect is usually seen on camera as an unsure performance.

When a person's fellow worker, employee or employer appears on camera, the tendency is for the onlooker to be critical. The onlooker has not had the practice at being tactful that director and crew have developed. The director should never risk exposing the nonprofessional performer to this sort of treatment.

When taping with monitoring equipment outside someone's office, the director should minimize onlookers' access to the screen and audio playback. Working with office personnel and the tape operator, the director should create a form of audio-visual privacy for the performer and maintain that privacy.

To ensure privacy, the director may have audio monitored on headphones instead of speakers and restrict the viewing monitor outside the office to the smallest size possible. Line of sight to a monitor can be easily restricted without any overt exclusion of interested persons. In a studio environment, access to the control room can be limited to involved and required personnel. Crew members can also create a respectful and helpful environment by their own actions.

PREPARING THE TALENT

The most important point to bear in mind while working with nonprofessional talent during the pre-production process is that the more the talent is aware of what to expect, the more successful his performance is likely to be. The director who has been through the production process on many occasions before is in an excellent position to prepare the inexperienced performer for each step of the production process.

The director should outline each upcoming phase in the process, delineate what is expected of the talent and eliminate surprises. This will allow the talent to feel and appear successful each step of the way. The net result will be an increased sense of self-confidence on the performer's part. This is the key ingredient in the creation of a good performance by an inexperienced talent—confidence that he will succeed.

Enhancing Self-Confidence

To heighten the talent's self-confidence, the director should outline each step in the production process in a fashion that will be understood by the talent. He should avoid all industry jargon and shape his speech so that the inexperienced performer can understand him. This may take some shifting of gears on the director's part, since jargon is essential to directing a crew quickly.

The experienced director will have internalized a linguistic shorthand that allows him to communicate quickly with the crew with a minimum of words. When he works with professional talent, he also relies on key phrases (more jargon) to get the talent to understand his direction quickly. But using the same jargon with the inexperienced talent will create a losing scenario.

The talent may act as if he understands statements, which he doesn't, or the talent will fail to carry out directions that he did not understand. This failure in performance will reduce his self-esteem and undermine his self-confidence.

Consequently, the director should consciously build the talent's vocabulary step by step and watch carefully for signs that the talent might not comprehend any given statement. This may require a particular alertness, since most working environments don't encourage the display of one's ignorance, and the novice performer often acts as if he understands discussions when he doesn't. It is the director's responsibility to educate the performer if the performer needs it, and to do so in a supportive fashion that builds his confidence, rather than erodes it.

Explain Production Techniques

Particular production techniques should be explained early in the process. Often, it will be necessary to explain them further as well. For example, if multiple cameras are used, it will help to explain camera direction and interaction with the floor director at the initial meeting, then reiterate those directions at the time of taping. This approach reduces the amount of totally new information that the talent is confronted with at any given point.

If single-camera technique is employed, a similar explanation may be in order. Otherwise, the talent may have an extremely difficult time adjusting to working out of sequence—taping his opening and closing remarks first, then doing isolated segments in the sequence that is easiest to light. Knowing that such an approach will be used allows the talent to prepare mentally to work in modular fashion.

The Pre-Production Follow-up

Since preparation is so important, it may be helpful to use a checklist for the topics to be covered in the course of pre-production meetings. Cover those topics and then send a follow-up memo prior to production that confirms the specifics. Following are some points which should be covered in print via a memo, note or follow-up letter, depending upon the director's exact relationship to the talent:

- Define the exact time and place of a taping session (a lost and late talent becomes a harried, poor performer).
- If the taping is at the talent's office, specify how much time is needed to set up, light and style the location. Suggest alternative work space for the person whose office is intruded upon.
- If makeup is provided, reiterate that fact to minimize any surprise or embarrassment on the part of an unprepared employee.
- Suggest wardrobe options based on an assessment of the performer's personality, style and the taping environment. Suggest specific colors with an eye toward their harmonizing with set or location as well as the talent's skin tone and hair coloring.
- Restate clearly whether notes are allowed, cue cards are provided or teleprompter will be present.
- If breaks or mealtimes are planned, state clearly those times as estimated so the talent will not feel cut off from opportunities to stay in touch with his business requirements. (If the taping is at a studio or on location, note that telephone access will be provided.)
- Clearly list what props or materials the talent is expected to bring to the session.
- Provide an honest estimate of the amount of time required. Include rehearsal time so that the talent expects it.
- Close the correspondence with an invitation for questions and provide your telephone number so that the talent can contact you right up to the last minute.

Create a Sense of Importance

Bear in mind that the taping process may be of minor importance to the nonprofessional who may have a multitude of other responsibilities. The follow-up in print gives him a chance to review material at his leisure between other tasks and may bring concerns to mind. It will serve to make the process more real, and often will create new questions as he begins to visualize the experience. Consequently, providing access to the director to answer new questions can be a great help.

The Production Process

Since a director is most successful with a nonprofessional talent when he can pay attention to the talent's personal feelings as well as performance, it is especially important that the director do his homework by organizing the production so that he can give priority to the talent.

Delegation

The director should pick a crew to whom he can delegate setup tasks, so that he doesn't have to oversee every detail. It helps as well to have thorough pre-production meetings so that the crew will execute the director's wishes while the director is paying close attention to the talent in rehearsal. Whenever possible, the director should avoid splitting his attention between the talent performance and other production concerns, such as lighting and continuity.

The director should make sure he is informed when the talent arrives. Then he can introduce the talent to appropriate crew members with whom the talent will interact. The makeup artist and floor director may have the most contact and the talent should be made to feel comfortable with them.

Create Crew Identities

I have found that it makes a noticeable difference to introduce crew members by name and title and then immediately explain how they will be working with the performer. For example, "This is Billy the prompter operator. She will work with you to run through your script on the prompter and make sure it's set up the way you want it. She can make any changes you need right now." This approach introduces the crew as a support team for the talent, rather than as a bunch of specialists who seem to be extensions of their equipment.

Eye Contact

At the start of the taping, the director should give the talent specific directions. For example, tell the talent exactly where to direct his eye contact ("right here, into this glass eyeball, that's where the audience will feel that you're looking right at them").

The talent should always be directed to hold his gaze to the camera at the end of a segment if that is what the director expects. This detail is easy to overlook and I sometimes ask the crew to remind me if I forget. (Many performances have been ruined by the performer who runs through a long speech perfectly to the very last word, then immediately turns off–camera without missing a beat to ask "How'd I do?").

Establish a Timetable

At the start of a production the director should establish a cutoff time and adhere to it. He should let the talent know that the production will stop at a given time. Without a cutoff time, the talent tends to constantly wonder, "How much longer will all this take?" This detracts from the performance. The same problem occurs if the director runs past an established cutoff time, as the talent constantly wonders when he will actually be finished.

The director and crew should show respect for the talent's time. A too-casual atmosphere will indicate to the talent that the process is not to be taken seriously. The crew and director should project a professional attitude by being prompt and ready to roll when the talent is ready. This indicates that all are trying to get the best performance possible by making the best use of everyone's time.

MAINTAINING TRUST

The Importance of Being Honest

Honesty should prevail when working with talent. One theory suggests that the director can get a good performance from a novice by "fooling" him into performing while he thinks the tape isn't rolling. This theory is based on the idea that inexperienced performers tense up when they know that they are being taped, but give good performances when they think they are only rehearsing. While this approach may occasionally result in a good performance being taped, the risks are simply too great to make it worthwhile, ethical questions aside.

If a performer feels hoodwinked by crew and director, he is unlikely to listen to their suggestions. A better way to address the problem of the talent tensing up is to keep the call to roll tape, the countdown to the studio and the call for action as low-key as possible—i.e., to use a normal conversational tone rather than a "countdown to blastoff" approach. The less tension projected by the director in this situation, the less the talent will perceive a need to tense up.

Immediate Feedback

It is equally important to give direct and immediate feedback to the talent after each take. Remember that the talent may have just done something for the first time that involves putting his ego and self-esteem at risk in front of a room full of strangers. He will need some assurance that he at least did not appear utterly foolish. The director should respond immediately to the performance with at least a "thank you, relax while we review tape." It is even better to compliment the talent on some point and indicate whether there will be another take.

The Stepped Improvement Approach: Win-Win in Practice

I have found it most useful to have my own game plan for shaping performance, particularly when many corrections are required. For example, I may know, based on rehearsal, that I will have to work on eye contact, gestures, pacing and other elements. In such a case, I may remind the talent to make good eye contact prior to the first take. Then, after take one, I can compliment him on his eye contact, giving him positive reinforcement, and suggest that we work on gestures in the next take. After take two, I can compliment him on the improvement and refine even further on the next take, and so on.

This approach allows the talent to succeed with each take. It forms the basis of a win-win strategy and works much better than overwhelming him with everything that's wrong after take one.

COMMON PROBLEMS

The Stiff or Oratorical Speaker

Many times, a person who has proven a successful speaker before large groups, is asked to create that experience on videotape. The speaker may have developed patterns that work well with a large group, but which don't translate to the small screen. Those patterns may include broad gestures, a well-projected delivery, and a generally broad style that plays well to the large group as part of an in-person presentation.

However, the same techniques captured on tape will often seem stylized and border on self-parody, since the camera in close-up gives the individual viewer a much more personal viewpoint than can be had by members of a large live crowd. At the same time, the speaker has usually had great success with the techniques that he's used in the past and will be reluctant to let go of the elements in which he has confidence. The successful large group speaker is also someone who enjoys "working a crowd" and wants to appear in control in public. Trying to shape this person's performance can be a serious exercise in diplomacy for the director.

Stressing Intimacy

I have found that it is important to direct such a speaker on an intimate level—to speak to him quietly out of earshot of the crew and any onlookers. In this one-to-one environment, I will explain that "television is a very intimate medium" and ask him to imagine that he is talking to just one person. "The camera is on a close-up in this section, and it will work best for you to picture that you are at a desk with just one particular person sitting in the room with you." I will also explain that the lavalier microphone is highly sensitive and that, for technical reasons, it will work best if the talent tones down his delivery to accommodate this feature of the medium.

The approach avoids criticizing someone's patented delivery. Any problems should be attributed to the new medium in which he is working, not to the tried and true techniques with which he is comfortable. A criticism directed at the technique which the speaker knows to be successful will fall on deaf ears, but new knowledge about the delivery mechanism will most likely be met with receptiveness.

The Distraction Factor

All of us become blind to the things in our surroundings that we encounter every day. Things which are very new and unique to the outsider may be invisible to the persons who work with them every day. Director and crew are no different. They work with powerful lights that create a hot studio or location environment. They step around cables and create a hushed atmosphere during a taping so that the audio track is clean. All their attention focuses on what is before the camera, so they can critique each performance quickly.

To the novice performer this is a highly artificial environment. At times it seems frantic as people rush around manipulating light stands, flags, cables and props. At other times it becomes deadly quiet as all attention is focused on the performer.

Helping to Maintain Concentration

In such an atmosphere, the performer's concentration, which is critical to a good performance, may be easily broken by small distractions. Sometimes the performer may even use the distractions as excuses for his poor performance.

It is very important for the director to spot this pattern and intervene. The director should pay close attention to the talent's performance and note the environment off-camera that may contribute to the talent's lack of concentration or momentary lapses. Then the director should—discreetly—eliminate any distraction which is not essential to the production.

Motivating the Underachiever

Some speakers will have reached a taping session through no desire of their own: the boss may have decreed that they would appear. Or they may have originally felt enthusiastic, but as the realities of putting their self-image on the line developed, they may have lost their enthusiasm.

This sort of person may simply be trying to endure the ordeal until it is over. He may, in effect, be unwilling to try very hard, since it is more difficult to accept failure if it is the result of a serious personal effort. Or, he may be unconsciously trying to wear down the crew until he can simply go home and forget about the experience.

The "Shame" Technique

This situation presents a serious directing challenge. Advice will be tolerated by the performer, but not seriously put to use, since the performer is not motivated. In such instances I will employ the "shame method." This is a technique in which the performer is treated with extreme solicitude.

If he even hints at a dry mouth, I will ask him if he would like a drink of water and immediately send the floor director to get one. I will look up at the lighting grid from his eye-line and ask if a particular light might be blinding him, and without waiting for a response, send a gaffer up to trim a barn door. Some of these may not even be real or bothersome distractions, but I will mobilize several people, stop all other activity, and make the underachieving talent aware that an entire crew is attending to him.

This approach usually has the effect of shaming the talent into making a greater effort, since so much attention is being provided in an effort to help him. Even if he doesn't feel that there is a better performance within his capabilities, he will typically try much harder on a series of takes after this technique is initiated. (This is particularly true in the case of the person who acquiesced to a superior's wishes, since this is the type of person who is most directly influenced by pressure from those around him.)

Making It Easier on the "Self-Criticizer"

Many persons try too hard before the camera. They have the television equivalent of stage fright and their own nervousness causes them to be overly critical of their mistakes. Such a person may also be easily intimidated by the taping environment and by the professionalism of the director and crew. He becomes painfully aware of how unprofessional his own performance is.

These sorts of persons can work themselves into an almost hypertense situation in which each small failure becomes devastating. They tend to lose their ability to realize what is important to focus on and what should be ignored. The director can and should intervene in a number of ways to break the cycle that is generated by the performer's failures.

When I notice that the talent has reached a point such as this, and he has been on camera for many takes, I will ask that the lights be doused and the crew take a break. This will be done in a fashion that indicates that this is a normal course of events, not the direct result of the talent's failure. This will allow the talent to re-assess the situation and regain perspective. It will also allow me to talk quietly with the talent and provide reassurance that multiple takes are quite normal.

Keeping a Perspective

I will also relate that I have the easy job, when compared to the talent's. I am working in an air-conditioned control room (or the cool unlit part of a location). I often indicate that we have plenty of time and that videotape is inexpensive. Any conversation that eases the tension and makes light of the situation for the hyper-tense talent will begin to break his losing cycle.

Admitting Errors

A further technique to use is to admit mistakes of your own. After a take involving a camera move or a switch, if the talent's performance has been less than perfect and the talent has been particularly hard on himself, I may tell him how he might improve and also note that I wanted to do another take for my own reasons. I either called a camera late, or a zoom wasn't at the right time. This lets the talent know that he is not the only one who makes mistakes. It also humanizes the director and crew and can serve to break any tension based on too much self-criticism.

Creating a Reassuring Atmosphere

Probably the most difficult role for any director to play occurs when he is running behind schedule and feeling pressure while working with an inexperienced performer. In the here and now, the director can only get a good performance from the talent if he internalizes his own tension and appears calm and collected as crises occur. It rarely helps the talent's performance for the director to project his own tension. This creates another distraction factor and even raises questions in the talent's mind as to whether the situation is out of control and possibly no good final product will come from it.

The best approach that I have found is to reassure the talent outwardly that this is not an unusual number of takes and that the crew and director are paid to make the talent look good.

CONCLUSION

The Larger Viewpoint

On the larger scale, the director may wish to make a key decision before the production even begins. He might look at the framework in which the performance will be taking place and ask himself, "Who is this person within the client base?," keeping in mind that the inexperienced talent's perception of the video production process will be greatly shaped by his direct experience in front of the camera. When the talent is the client, an unpleasant experience before the camera will likely diminish the chances of there being more video done by the client's group or department.

If several projects are contemplated, and the talent will have additional chances to appear and improve, the director is working in yet a different framework—one in which he may see the first taping as the start of a director-talent relationship. He may plan a progressive development of the talent's performance level and settle for an adequate performance the first time out. This could be premised on an understanding that the particular video is of lesser importance than future tapings and that the talent will improve with a gain in self-confidence.

The Performer's Perspective

Overall, the director should weigh the question of how far he wants to push this particular person, during this one taping, to get a performance on tape that is up to a particular standard. This factor should be weighed against the question of how much he might be willing to risk the person's self-esteem, and how unpleasant he might be willing to make the overall experience in order to get the best performance on tape. This is especially important when the performer is also the client.

The Director's Perspective

Since the talent is not a professional actor for whom "what's on tape" is of ultimate importance, the director should maintain his own perspective throughout the process and know "when to quit," when to accept a level of performance. Knowing this somewhat ahead of time will allow the director to accept that level of performance in a positive fashion, which allows the talent to feel good about himself. The other scenario that is enacted too often is one in which the director makes it obvious that he is accepting the last take only because he has run out of time, patience or the hope for improvement.

Although this latter approach leaves the director's ego intact, it does little for his relationship with the talent outside the taping process. Ultimately, the director should remain very clear about the differences between working with professionals, whose lives are structured around creating a good performance, and nonprofessionals whose performance is only a blip in a professional life that has its own rewards. An entirely different set of expectations comes into play when working with nonprofessional performers than when working with professional actors.

9 Directing Professional Talent

The director will have a very different working relationship with professional talent than with nonprofessional performers. One major difference is that the director has the opportunity to audition professional performers until he finds the ones he feels are most suited to the roles in the script with which he is working.

SEEKING THE IDEAL PERFORMANCE

Prior to the audition, the director examines the script to visualize possible interpretations for each character. The idealized performances that he creates in his mind are his guide for the audition process. Each performer whom he auditions may fall short on some points, meet his expectations on others, and bring new dimensions to the role in other aspects. The director leaves the audition with a new range for "ideal performance" in mind, one that has been tempered by the real performances.

He then makes a casting decision which measures what each actor may bring to the role. The chosen actor knows that the director selected him for a role. Something in the talent's background and performance inspired the director to give him a vote of confidence at the outset of the production. A basic trust in the actor's ability has been implied, and it allows director and talent to deal more directly with the script and its demands, as well as with each other's expectations.

FOCUS ON CHARACTER INTERPRETATION

So, instead of concentrating on the psychology of the actor (as the director must when dealing with nonprofessional performers), the director can focus on the psychology of the character whom the actor plays. This involves examining the script a second time, after casting, and re-visualizing the character based on the actor cast in the role, as well as the other talent chosen for the roles which will interact with the actor. The casting of other roles in turn affects how each role might be played.

When working with professional actors, the director can focus on a more candid communication of ideas. Since the director and the talent have worked in the field of performance before, they can take advantage of common experiences and language to communicate with each other. They can enjoy a much more effective working relationship with the opportunity to add more nuance to the performance.

The director must be able to communicate the portrayal of the character that he has visualized. This is typically accomplished through several phases of production:

- the audition dialogue
- pre-rehearsal script conferences
- pre-performance rehearsals
- on-site rehearsals
- the performance itself

In the initial audition, the actor gains a basic sense of what the director has in mind for a specific character. The director then sees other interpretations from other actors and he can rethink the role with the new input. His original interpretation may not be the one that he later requests.

Script Conferences

The next point in the production at which the director can convey his interpretation of the role is during the script conference. This typically follows shortly after the casting decision has been made.

At this time, the director should give the basic parameters for a character to the actor and guide the actor's thoughts as the actor begins to learn his character's lines. The director should be specific enough so that the actor doesn't develop the character portrayal in a fashion which the director cannot include in the production— but at the same time, the director should leave open those areas in which he would like the actor to explore options.

Particularly when working in a union environment, the director should remember that the script conference is just that—a conference, not a rehearsal. The script conference is simply a forum for actor and director to develop a common understanding of a character and specific production logistics without actually reading lines from the script. (Union contracts are quite specific on this point.) It can also be an opportunity for actors to meet their opposing players, so that they can read their scripts with a sense of what interactive opportunities may exist, and a chance to tailor their interpretations to accommodate the other performers' characteristics.

Marking the Script

Even when the director must rehearse talent at their first meeting, the director should use the conference concept prior to rehearsal so that actors begin rehearsing with a common understanding. At the conference, the director should comment on the overall purpose of the script, highlighting the elements which he feels are most important and indicating the type and level of emotion at each phase of the script. One technique that is useful is to mark the script in terms of intensity, with 1 standing for least intense, 2 for more intense, 3 for even more intense and 4 for the points of greatest intensity. A color coding system can be used to the same effect.

This allows the director and actors to visualize a rise and fall in the script by glancing through the pages without reading lines. It gives them a common structural interpretation of the script prior to rehearsal and is a much more efficient use of time than a cold reading.

TECHNIQUES FOR COMMUNICATING WITH TALENT

During rehearsals, the director can shape performance without the time demands, distractions and technical concerns that mushroom during the production itself. In order to use this time well, the director should develop his ability to communicate his thoughts to the talent. There are several conventions which can be utilized, since acting for video is simply the latest in a historical progression of dramatic forms. Director-actor dialogue has been refined through centuries of stage drama and decades of film drama.

The Demonstration Approach

In this approach the director will choose to act out the role as a way of showing the actor what is in his mind. This approach may work very well if the director has innate acting skill or has worked as an actor, and if the director has a good sense of self-awareness. (But if the director is not an accomplished actor, he may think he is presenting the performance that he desires when he is not. In fact, his performance could mislead the talent.) The demonstration approach tends to work best when it is used to define the reading of specific words in a script—for example, when working with spokespersons or narrators, rather than with dramatic characterizations.

Another drawback to this method is that it limits the opportunities for the actor to broaden the director's horizons. The actor who simply mimics a director's demonstration will lose many of the unique aspects that would arise from his own interpretation based on the director's verbal direction. Inexperienced or timid actors, in particular, will tend to mimic the director's actions rather than take risks with their own intuition. This will limit the scope of role interpretations.

Motivation and Subtext

Motivation and subtext (discussed later in this chapter) are two conceptual tools that enable the director to communicate his conception of a performance without unduly limiting the actor's options. The first is the concept of background motivation, the second is the idea of a subtext. These are similar concepts but they differ in ways that can prove useful to the director who wants to create subtle performance shadings.

Motivation Consistency

The concept of motivation simply addresses the fact that, in any script, the author's characters may move and talk for reasons which are not directly evident from a reading of the script as written. The director allows his imagination to fill in the blanks and create reasons for actions—reasons that are not contained within the script itself. The process is not an arbitrary one, but is based on the director's thorough reading and consistent interpretation of the script. The director must create a consistency in the characters' motivation so that they appear to the audience as believable characters. Actions should be justified; they should not "come out of left field."

In its most basic form, motivation simply relates to the placement of furniture, doorways, windows, etc. During rehearsals, this might be termed "unseen architecture" or "scenic hardware." For example, the director may call for a character to cross the frame from one side to the other in the middle of his dialogue to add movement to the action. The actor playing the role asks "What's my motivation?," his way of trying to develop a style for the movement that fits within the set of actions and patterns of voice and motion that make up his character. The actor also wants to maintain a consistency of character.

Underlying Motivation

The director's response should indicate a form of motivation that is within the bounds of the character as he sees him played. His response could be, "Your motivation is your desire to create more distance between yourself and the other character," or "to get closer to the window, since you're thinking of how you might get out of the room."

This gives the actor more information relating to general feeling and potential types and styles of movement that might be appropriate. It is also through the discussion of these subtleties that director and actor will arrive at a deeper common understanding of a character as a whole, which, in turn, enriches the actor's overall portrayal.

Talent Embellishments

These discussions also give actors a base from which they can suggest embellishments to the character's actions, to add bits of "business." In the above example, the actor, now understanding that his character wants to make an escape from the situation, might suggest that his character subtly pull his jacket more snugly about himself, as consistent with someone who is thinking about the shock of entering the cold night air. The director can consider this and give the actor feedback regarding how thoughtful or methodical the character is and whether that action really fits the director's conception of the character.

Physical Staging

This give and take also influences the actual staging of the physical aspects of a production. The director may have complete control over many aspects of the setting of a production. He may be able to design and build certain stage elements. Or he can dictate what locations should be considered and what elements are necessary to each location. During script discussions, actors can make suggestions which the director can then incorporate into the stage or location design to give the actors the opportunity to add their personal creative touches to the characters.

The Director's Overview

However, through all this, the director must maintain the overall view. He is the directing force who judges which aspects are appropriate to the characterization and which are not. One actor may suggest a very dramatic movement which appears at first glance to embellish his performance and add interest to a particular moment in the script. It is up to the director to decide whether that suggestion would call too much attention to the one character at that time and distract from more primary actions. Often, the director must mediate when various actors present differing views of the relative importance of their respective characters and their actions.

Motivation extends beyond the simple movement about a location or stage, however. It is also one of the ways in which the director and talent reach a common understanding regarding a specific character so that the talent can choose the appropriate tone of voice, body posture and attitude to accompany particular dialogue. It allows the director to specify the "why" of a character's actions so that the actor or actress can internalize more than the words on a page. The talent will use the director's choice of motivation to guide his performance.

Alternative Motivating Factors

For example, in a given scene an actress may be called upon to play the role of an executive who must discharge another employee. The script may specify that she greet the employee in an outer office, guide him to her interior office, offer him a beverage and then proceed to tell him that he is fired. At a particular point, the script may call for her to interrupt the proceeding to close a door. The director may specify that her motivation for the action is to provide privacy for the employee. This would suggest that she carry that solicitude through to her actions during the scene, particularly in the way in which she addresses the employee as she closes the door.

The director might, on the other hand, suggest that her motivation is simply to close the door so that she is able to hear the employee without the distraction of outside conversation. This would suggest that her attitude be more offhand; that she even throw a look to the area outside the office and perhaps even act distracted prior to the moment of closing the door. The director can use the motivational concept as a shorthand that allows the actress a range of interpretation for a singular action.

Subtext

A second concept that can prove helpful in elucidating a script is the concept of "subtext." This is a larger view of the unseen and unspoken aspects of a script interpretation. A motivational concept may answer the question "Why this particular action?" and a subtext answers "What's really going on?" in a scene.

As an example, let's assume a particular subtext for the scene described above. The director may specify that the subtext of the scene is that there has been a long-standing feud between the executive and the employee about to be terminated. Further, the executive would be deriving pleasure from the act of firing a potential rival. This would definitely color the actress' performance and allow her to add shades of meaning to her lines that would not suggest themselves without that particular subtext. She would appear in control and possibly eager in her task.

By contrast, the same words may be applied with a different subtext to provide an entirely different style of performance. Suppose the actress were told that the subtext was that the employee was her supportive friend, but that he had violated a technical rule in the company and therefore had been ordered dismissed despite her objections. This would provide an entirely different guideline for her general bearing and interpretation. The same words might be used to project regret, reluctance, genuine sorrow and a sense of loss of control of the situation.

The director should examine any dramatic script for the areas in which a sub-text may be operative and where specific motivation might be called for. This will allow him to approach a script conference and subsequent rehearsals with a clear sense of the key information that he would wish the talent to understand, and it will lead to rehearsals that result in more focused performances, more quickly.

THE DIRECTOR'S CHECKLIST: SCRIPT FINALIZATION

To assist the talent in their review of scripts, the director should have an internal checklist of important areas.

Provide Finalized Scripts Promptly

First, he should make certain that all actors and actresses are given finalized scripts at the earliest possible convenience. (The scripts should be finalized because nothing is worse for an actor or actress than to have to quickly "unlearn" lines that were changed after they had been memorized. In addition, it is harder for the actor to give a good performance when he has to remember what has been changed and what hasn't.)

A second reason for early script delivery is to give actors enough time to study their scripts, particularly because many hold additional jobs.

Agency Script Delivery

The director should be particularly wary of script delivery that is handled by a talent agency. Smaller nonbroadcast and non-feature productions that involve only one or two players for only one or two days may not be given the same care and attention as larger-scale productions which generate greater billings for an agency, and some agencies that do primarily modeling may not be conscientious about prompt script delivery. The director must allow for a time lag between agency receipt and delivery of a script unless he knows the agency well.

Mark Scripts for Memory Requirements

It will be a great help to the talent for the director to indicate on a script what portions need memorization, which areas will be on cue cards or teleprompter, and which areas will be read as voice-over. This lets the talent concentrate on memorizing only the necessary parts of the script, becoming very familiar with the prompted parts, and simply reviewing the voice-over elements. (The director can, of course, cover his options by indicating as "memorized" any area that he thinks might need to be memorized.) This approach can pay great dividends, since the director gets more focused effort from the talent where it is required.

Prioritize via Shooting Sequence

If the director knows that the talent is pressed for preparation time, he may wish to indicate the day's shooting sequence. This will allow the talent to be prepared for the first scenes and to spend time between setups to prepare for succeeding scenes. The only risk is that weather or some unforeseen factor may dictate a change in the shooting schedule. But, again, the director who knows his shooting crew and production circumstances can usually predict the degree of actual risk involved.

REHEARSALS

Rehearsals provide a very creative time for the director, since it is his first chance to see the actual character interaction with real bodies in real time. Many new ideas regarding staging, timing, costuming, props and background will occur to him as the first rehearsals unfold. To make best use of the time of all the players, the director should keep in mind that his role has a much larger view than that of the actors and should not look for approval from them for every creative decision.

In other words, he should remain clear throughout the rehearsal process that he has his own larger agenda, that he is responsible for guiding talent performance; but he can and should be refining the entire production during this time as well. To involve the talent in many of those considerations is distracting for the talent and can cause the director to lose his own focus. This can be particularly true when an inexperienced director works with experienced talent, since his tendency is to look to them for approval for his decisions. As comforting as that approval might be, it creates a confusing climate for director and talent alike.

Schedule Rehearsal Time Appropriately

The director should schedule his rehearsals close enough to the final production date so that the work done in rehearsal is not forgotten or intruded on by other roles. Keeping in mind that many performers are involved in other jobs (and often, ongoing plays in production) the director must realize that other characters and scenes may be played by his talent after his final rehearsal and before the day of performance. The involvement of his talent with other productions can result in an unintentional reshaping of areas of portrayal that he might have worked hard to develop with the talent. The director can get a good sense of the level of concentration and attention of each actor by casually asking about what he is doing outside the production, and showing personal interest in his other activities.

Get to Know the Talent

This allows the director to gauge the number of distractions that will intrude on that particular performer and, correspondingly, to adjust how much direction he provides in each step of the production. For example, if one performer is involved in a play that runs nightly, the director may choose to focus that performer on character assessment and line memorization until the performance itself. He may use the other performers, who are better able to prepare, to flesh out details off which the distracted actor can play. Obviously, these details should not involve any major blocking or timing cues, or a disparity will be created.

Probing for outside activities and personal career information also lets the talent know that the director is sensitive to them as individuals (without necessarily creating a forum for excuses for bad preparation). The director who shows genuine concern for his talent will find that the talent feels more committed to giving the director a better performance. This approach can also lead the talent to be more open with their personal observations while respecting the director's role.

There is another benefit from developing relationships with the actors in a production. By discussing career paths and goals, the director can encourage the talent to make more creative suggestions, ones that may seem high risk with low rewards within the immediate production's context, but which might allow director and talent a growth opportunity within the framework of their own career development. Each production exists within the larger scope of the actor's and the director's career and can provide an opportunity for either or both of them to try something new.

A Personal Example

For example, while directing a United Way campaign program, I once cast an actor who had developed a classic "nerd" character—a Milquetoast-type with glasses—for several years and had played that type of character quite well in many commercials. The writer had asked for this stereotype in the script and the client had come to expect just this portrayal. However, during the first rehearsal, the same actor mentioned that he had become so worried about being "typed" as a nerd for the rest of his career that he had recently acquired contact lenses to open up his casting options. He politely suggested that he could play the character as less of a stereotype, since he could play the role without glasses.

I reviewed the script and found that it was not only possible, but could strengthen the program by making the character more of an "Everyman" with whom the audience might begin to identify. In the larger picture of the program's objective, this would actually work better than relying on an easy stereotype. I convinced the client of this viewpoint and, needless to say, the actor made a great effort to create an effective performance.

We had a very good relationship throughout a demanding production schedule, partly because the actor realized that the director was open to more than a tried-and-true characterization. But the fact that the director had been interested in what else he could personally bring to the production meant that his efforts as an actor would be better appreciated.

The Director's Notes

Small nuances of behavior may be forgotten in a short time after a single rehearsal. No matter how close in time he schedules rehearsals to performance dates, the director should still limit the level of detail that he shapes in rehearsal, working instead with the larger scope of the talent's performance. He should keep notes on his script regarding particular details that he wants to work on later in production, instead of encumbering his talent with too much detail at the time of the rehearsal. This time is best spent developing the overall talent performance and creating a fluid interaction among characters.

This discipline takes practice, and it is one of the many "simultasks" that the director is called upon to perform. He needs to continue his active presence in the rehearsal action while, at the same time, making side notes regarding future modification of the talent's performance. This may seem an awkward technique at first attempt, but it pays great dividends in terms of a well-crafted overall production. It allows the talent to maintain the flow of the rehearsal performance, and lets the director keep his eye on details as well.

Limiting the Level of Detail

The director should let the talent know what the basic layout for the scene's furniture and props is, so that their general movements will fit the scene, but he does not need to tell them the exact placement of every item. In rehearsal, he should not try to get the talent to move just the right number of inches to go from an unseen chair to an unseen door. This tends to lead to confusion and can get in the way of working on the more important elements in a scene. The director's task in rehearsal is to make use of the time in a way that will pay the greatest benefits during the performance.

Ranking Scene Elements

Part of this equation is the director's responsibility to set value on the relative importance of parts of scenes and to allocate the time needed and level of preparation required. He should examine the purpose of each part of the script and decide where he will need the strongest performances, where he might accept a lesser performance if necessary, and so on. In doing so, he is looking at a script in terms of "dramatic intent"—the purposes of the scenes as defined by the author, himself and perhaps even the client. This is a different process than examining the script in terms of motivation or subtext.

For example, the dramatic intent of an opening scene may be to establish the basic characters in the drama, create an awareness of the characters' mutual relationships, and define any tensions, conflicts or bases for the action about to unfold. A later scene which serves to carry the action forward without as much exposition may be of a lower priority for the director in rehearsal. Further, the director might choose to rank scenes in terms of their critical impact on the actors themselves. If certain key scenes are worked out fully, they may define the characters for the actors in such a way that other scenes easily fall into place.

Knowing these facts allows the director to allocate his use of rehearsal time better than if he were simply approaching the script as a whole entity. Instead of trying to create an evenly perfect performance in the rehearsal process, he should concentrate on developing the key elements. This activity will, in turn, give the actors the clues needed to refine their interpretations of the lesser developed parts of the script.

Experimentation Opportunities

The director can also use the rehearsal as an opportunity to experiment with different ways of playing the same scene. In many scripts, the author leaves a wide range of interpretation available to the director. Often, the only way to know whether a particular scene will work in a different fashion is to actually play it out with the actors who have been cast. Once the director is satisfied that he has covered the key aspects of a script in the rehearsal, he may wish to continue to experiment.

Reviewing for Consistency

However, when changing any key scene's tone or modifying specific lines, the director may wish to first make it work in the small view, and then run the scene from the top, or play several scenes out, to make sure that he has not created a dramatic inconsistency. He should never make major changes to the scenario without a reading check prior to concluding the rehearsal. Otherwise, he may be in for a very unpleasant surprise during the final production.

The newly developed interpretation may seem drastically out of place when viewed within the program as a whole during production, necessitating immediate rethinking on site. Or, worse yet, the director may be shooting out of sequence in such a fashion that he does not discover the inconsistency until the edit stage. At this point, he will have to reshoot, or risk compromising the overall production.

THE PRODUCTION

During rehearsal, the director should discuss the logistics of the production days so that the talent have a good sense of the pacing and structure of the shoot. This will allow them to prepare themselves for the process as well as for the role.

Production Style Considerations

A sequentially produced *multi-camera* shoot will have a relatively slow start-up period with longer rehearsal segments. Camera and talent blocking will have to be worked out. Lighting and audio cues will need to be refined. Then, large blocks of script are shot in real time, with the talent having to remember specific dramatic and camera direction through longer parts of a script.

There will then be longer periods of downtime for resetting camera angles, changing and checking backgrounds, props, and set elements, during which the talent will review the next large block of script. Then the longer rehearsal and taping process will begin again. The length of breaks between takes will test the talent's ability to remain "up" and recapture their character quickly once the new take begins.

A program produced using *single-camera* film-style techniques will be more segmented. It is usually shot in a different order than the script. It will also involve shorter script segments per setup and will often require less setup time between takes. This creates an entirely different set of demands on the talent than most multi-camera productions.

Actors and actresses will be required to get in and out of character many more times per day. They will have to create one mood, jump forward in the script and change that mood drastically, then jump back in the script and return to the original mood—all for the convenience of the staging and lighting logistics that accompany out-of-sequence production techniques. This involves a different type of concentration and many performers have different ways of dealing with the challenge. Some never break character throughout the production day; others choose to break character totally as soon as they are off-camera.

The talent can give a more even overall performance if the director prepares them beforehand for his chosen production style. This allows the talent to take the larger view and to pace themselves in their own fashion so that they are able to give the consistent performance that is required by out-of-sequence staging.

I will often talk to the talent during rehearsal about some of my production strategies and my overall production design. For example, I might let them know that I have planned certain scenes at certain times because of daylight requirements or to make the best use of talent and crew energy. This lets the talent in on what I consider key elements of their performance, and how I see production concerns interacting with performance concerns. However, I will not do this without first making sure that they have a clear understanding of the script itself. Otherwise, this may tend to distract from their basic character development.

Client Considerations

If the director knows beforehand that there will be "active" clients present during the production, he should explain their roles to the talent.

Many novice clients may not understand that the director needs to be the sole source of direction for the talent. They may try to direct the talent because they are enthusiastic about getting a good performance and see things which they feel should be changed. Their timing may be poor, and their sensitivity to established roles may be minimal, but their good intentions and the fact that they are ultimately "paying the bill" may necessitate a careful handling of the situation.

As clients begin to assume the directorial role, the director may have to be very diplomatic in his assertion of the need to operate in a prescribed fashion. He may have to defer to some of their opinions, which may be correct, and override others without seeming to ignore their concerns as clients.

If the talent knows at the outset what relationships are involved, the talent will have a much easier time helping the director to handle the situation. Otherwise, actors can be abrupt and rude to people who seem ignorant and meddlesome, or, as they see the director defer to the client, they may become confused about "who's in charge here" and begin to assume that the director has abdicated his role.

The director who foresees potential problems and informs his talent about specific relationships can avoid these conflicts. For example, after the client suggests that the talent change a line, the informed talent might simply say, "That's really something only the director can decide," or in some other fashion politely indicate the normal working relationship. He might indicate that only the director works with the talent and the client should address any concerns to the director.

Client Role Definition

Naturally, there are many production situations in which someone from the client-side has a legitimate role as a content expert or technical consultant and it is more efficient for him to speak directly to the talent. In such cases, the director can explain this to the talent and ask the talent to let him know if the advice becomes too distracting. (The effective director will usually sense this fact as soon as the talent does, but by letting the talent know beforehand the director will strengthen their working relationship.)

Sometimes there may be a fine line between allowing too many cooks into the kitchen versus creating a rigid atmosphere in which no one feels he dares speak up. A confident director who has come to know his client as well as his talent, who has a specific production plan and who is aware of what is necessary to each role (and what is not) is better able to strike the appropriate balance.

Communicating Production Expectations

Even with professional talent, the director needs to be concerned with varying levels of expectations. When working with inexperienced talent, stage actors who are unfamiliar with the rigors of multiple takes for an edited production, or "personalities" or news persons who are used to single takes or live performances only, the director should make sure that they understand his expectations for acceptable performance and the realistic logistics required for the production. The talent should be aware of how long the shoot might last and whether there are options for rescheduling part of the production. Otherwise, the director may find himself pushing the talent past their expectations, and this may result in a very unpleasant and/or unproductive working atmosphere.

I once assisted a director who had chosen a local news anchor to host a corporate production to be shot in a single day on a soundstage. This was her first non-news performance, but she was quite good as a spokesperson. Unfortunately, the set graphics were unusable as delivered, had to be reworked on set, and this created a two-hour delay. She seemed to take no interest in the problem at first, but, as the production ran into late afternoon, she informed the director that she had opera tickets that evening and had no intention of working past 5:00 p.m. Since she had done only news, she was not accustomed to a different time frame and quality demand.

The production was hurried to accommodate her expectations and she was released by 5:30. However, the director had to give up several complex camera moves he had planned and did not get the production value that he had originally aimed for. If the talent had been informed at the outset that overtime might be involved (and in fact was a common expectation on single-day shoots) she would have scheduled her own life accordingly and been much more amenable to "getting it right."

Take Care of the Talent

At the time of the shoot, the director should make a conscious effort to maintain the relationship that he has been able to develop with the talent beforehand. With all the other concerns of production, continuity, camera, lighting and sound logistics, crew relationships and client concerns, it is easy for the director to be pulled away from the performance itself. Some basic considerations can help to maintain a closeness with the talent.

The Talent's "Personal Space"

The director should ensure that on any location, at any studio, the talent has a dedicated "green room," holding space, or personal area. This will ensure that the director can leave the set with the talent and discuss script and performance in a private area.

The director should remain aware that the talent has a major stake in looking professional and appearing in control in front of other working professionals. If specific problems need to be discussed, they may need to be addressed in private. Further, the talent who must wait for setups to be completed, while being surrounded by the setup activity, will be unfocused and out of character by the time the scene is ready. Many actors cannot resist an audience and will begin to play to the crew, leaving themselves less than rested and prepared when the scene is to begin. The director who deliberately controls his talent's surroundings will have much better success in controlling their performance.

Creating Director/Talent Intimacy

The simple act of meeting actors privately as they arrive at the production, reviewing their wardrobe and makeup and discussing the day, will create a more intimate and relaxed relationship between actor and director. This will, in turn, encourage better communication during the production day. A less obvious benefit is that the director shows the talent that they are respected for their skills and that their efforts will be valued. This is not a small gesture for persons who seek approval as most actors do.

Final Script Reviews

The initial meeting on set is also the time for the director to reassure the talent that the script they have is current, or, if there are changes, to provide them immediately so that the talent has time to learn or unlearn lines. If changes are sizable, the director should enlist someone else's aid to help the actors modify their scripts, so that he remains available to the crew. Actors become extremely dependent on their personal scripts at production time as they will have personal notations on them that are crucial to their performance. Without a current script, they can become quite unnerved.

I have found it extremely useful to copy my own script on a different color paper stock so that I can find it very quickly. Without my personal script, I am sometimes at a loss when I am asked "What's next?," and yet it's my job as director to constantly stay ahead of the production flow. It has proved a great time-saver to be able to spot my bright yellow script across the studio or at a distance on location.

The Importance of a "Warm-up" Period

To encourage good talent performance a director should allow for a warm-up period, no matter how pressed for time the production is. Actors need time to begin to interact with each other and recapture their rehearsal flow. They need to sense timing differences among themselves and work them out. The director will benefit greatly from a purely dramatic reading at the start of the day to tell him who is prepared (and to what level) and who is not. This will allow him to adjust his view of on-set rehearsals and focus his attention where it will do the most immediate good.

Initial Run-throughs

It is much easier for a director to get this sense in an initial run-through, which does not involve all the other production elements such as camera blocking, audio cues, etc. Even just a few lines read will provide valuable clues to the talent-direction task ahead. Without such run-throughs, the director risks having a logistic-driven or technically driven performance.

During initial run-throughs, the director should avoid quick criticisms of the talent, but should allow them to warm up to the situation. It will help to recall the phrases that were used to describe mood, motivation and subtext during rehearsal and use the same common language during on-set rehearsals so that the talent's memories are jogged. (This is particularly important when working with actors who have other employment or roles that tend to blur their script memory. The director should also remain aware of the effect of early call times on actors who do stage work at night and may wish to schedule key scenes later in the production day to compensate for a slow warm-up.)

Avoiding "Run-on" Performance

Stop points in scripts should be clearly indicated so that the talent do not continue with a scene past the necessary point. This creates the gradual running-down-hill syndrome as the talent begin to wonder how far they should go in the script and begin to become very unsure and halting in their performance, waiting for the director to stop them. Even when the talent are "on a roll" the director should call for the cut as originally planned. To act inconsistently with his initial requests will serve to confuse the talent.

Clarity of Direction

The same principle should extend to all directions. They should be clear to the talent, especially when an interpretation is being changed. The director should explain that he is asking for something different rather than criticizing someone's performance. This will ensure that the actors trust his judgment of their performance when he does ask them to try harder.

Conserving Talent Energy

The director should remain active, maintaining control of rehearsal and performance flow. If he becomes passive, the talent will tend to try different approaches until they get reactions from the director. This leads to a dilution of their energies and gradually to a less effective performances.

Using Talent Experience Effectively

When the director tells professional talent what the general framing of a shot is, they can modify the scale of their gestures appropriately. One of the pleasures of working with professional talent is that they have usually developed the ability to play to the camera and use screen space effectively. If a firm action match is required from one setup to another, most experienced actors need only to be told at the outset of a scene and they will provide consistent timing and movement that will edit easily. As the director comes to trust his crew and talent in these more technical aspects, he will have time to work on the subtler aspects of performance.

An experienced actor may often surprise a director and crew with his ability to spot technical stumbling blocks and provide solutions. This can be a great boon to a less-experienced crew and director.

An experienced stage actor will be able to tell when he "hits his marks" by whether his key lights feel warm enough as he passes through them. (An actor's key light is focused in the area where he will be positioned "on his mark.") An experienced actor knows that he or she can feel that warmer spot and not have to look down to find the tape "mark" on the floor, which detracts from a natural movement.

Tape Playback

One advantage of videotape is that any take can be replayed for analysis. This allows talent, crew and director to see where particular timing cues need work and to address them quickly. The disadvantage of tape playback is that the viewing process takes time away from working on the set and forces the talent to break character. The director who can spot many elements in real time and avoid replays will keep his actors at their peak and get more effective use of time from talent and crew alike.

Another time-saving and energy-focusing device is the use of uninterrupted rehearsal time to allow the talent to run a scene while the crew works out the lighting, camera and audio logistics. Since tape is inexpensive, some directors have come to blur the line between rehearsals and takes, putting rehearsals on tape in the hope that they may be adequate for use. However, this tends to give the talent a clouded sense of what the director is looking for in a performance.

The Director's Balancing Act

The effective director orchestrates rehearsals to provide minimal distraction of his talent while giving the crew a look at the flow of a scene from the technical side. Scenes are not put on tape until the director is satisfied with all elements. Keeping a clear distinction between rehearsals and actual takes will keep the talent focused on repeatable performances that can be refined. I am convinced that this leads to better performances on tape, since the talent moves to a keener edge after several rehearsals when they know that "this one's for real." This approach also reduces the amount of burnout, since rehearsals are less taxing on performers than potential takes.

The Importance of Feedback

After any take, the director should give his talent immediate feedback so that they can either continue preparing the scene at hand or relax and move on to the next scene in their minds. This may take conscious effort on the director's part if he is concerned about technical details such as potential background noises, camera bumps, and so on. But if he always addresses the talent first, before and after takes, he will find that his talent pays closer attention to his direction when he gives it.

Actors, like anyone else, appreciate positive reinforcement. If the director regularly compliments them on a good performance after each successful take, the talent is likely to settle into the role and continue to make the extra effort on subsequent takes. Genuine compliments, freely given, pay great dividends.

The director who knows that the talent has attempted a tricky or difficult scene should encourage the crew to join him in applauding the talent's successful attempt. The talent will get immediate positive feedback from an audience and will realize that the crew and director appreciate his skills and can recognize superior achievement. This not only creates a healthy working atmosphere, but usually results in stronger long-term working relationships.

Staying in Touch with the Talent

It is particularly important for a director not to be fooled by his talent's ability to look comfortable in difficult circumstances. On the simplest level, the director should note that professional actors are trained to look natural even when wearing a heavy suit and pancake makeup under very hot lights. He should remember that the talent may be much warmer and more tired than they appear and make a conscious effort to minimize their discomfort so that they can keep their concentration. Actors speaking lines will also tend to dehydrate quickly, which is why liquids other than coffee should always be readily available.

Take Number Strategy

When an actor is having a particularly difficult time with a scene and the take numbers have gotten very high, the director may consider rolling back the take number to Take 1. This can provide a subtle psychological boost to some actors, since they are no longer being reminded of their failures right before each take. In fact, I once reviewed several takes of an experienced actor with whom I had gotten to a very high take number. I could see clearly that, just as the slate announced each take, he cringed out of shame at his poor performance. This affected the beginning of each of his takes; he would start from a point of defeat each time. Once the slate was rolled back, he lost his self-criticism and contributed a good performance very quickly.

Experimentation

It can prove very rewarding for a director to pick certain scenes in a production for testing the talent's limits—to gauge when he has "covered" a scene on his own terms and then, schedule permitting, encourage the talent to "pull out the stops" and try another approach, perhaps a risky one. Over time, these experiences, combined with auditions, will give the director a better sense of the breadth and depth of the talent available to him in his area. This, in turn, will affect his view of how future scripts might be handled and will keep the director open to new approaches to characterization.

DIRECTING THE SPOKESPERSON

The previous discussion has concentrated on actors who play the roles of particular characters in a set of circumstances within a dramatic form. While many of the same principles apply to directing the spokesperson role, there are demands which are specific to the spokesperson role. These are discussed below.

A "Ready-Made Persona"

Dramatic characters are defined by the circumstances within a script or implied by a script. They have no life outside the dramatic form itself. Spokespersons bring a complete "persona," a character of their own, to a script. They are rarely affected by events within a script except in a most surface fashion.

The celebrity spokesperson is someone who is already familiar to an audience and who is felt to be an appropriate match to the product or service or company about which he speaks.

Non-celebrity spokesperson roles are slightly different. They require that an actor present a character to the audience who appears and sounds like a real person, yet is someone whom the audience hasn't met. They must command credibility based solely on their presence within the first few seconds of viewing. This requires the actor or actress to have previously developed a "character" with a consistent manner of speech, style of movement and general bearing—a character who can address an audience and speak convincingly about topics which the actor or actress may have no background in or knowledge of.

Choosing the Spokesperson

The director should cast for spokespersons with appropriate scripts, realizing the different set of skills that is required. Often a successful spokesperson will have a narrow dramatic range, but is a quick study of information and style. This allows him to absorb information about a topic as well as portray some of the aspects of a real person who might be familiar with the topic.

The director can assist a spokesperson's performance by giving him information about the company or product he is representing. This should be done in a broad stroke fashion. The director may include the basics of the company's history and current policies, plus a profile of the typical employee. Or he may explain a product and give the actor the background of typical users of the product.

Matching Spokesperson to Production

Many of the nuances of a given company's self-image are not obvious in a script. To be successful, the director needs to supplement the writer's work with his own awareness of "corporate-speak" particular to a given company, pronunciations specific to a profession or small details of interaction or movement—all of which can make a performance feel more real and appropriate for the particular company.

To do this, the director may have to research the company or product being represented, as well as the perceptions of the viewing audience toward the company or product. From this, the director and spokesperson will create a performance that becomes a vehicle for the acceptance of new ideas because it is based on elements familiar to the audience.

Rather than develop a character with depth, the director and actor will focus on wardrobe selection, exact pronunciation of key words and pacing of speech. The talent may have to learn large amounts of jargon and speak a technical language as if he does so every day. The director will have to guide the talent based on his observations of the appropriate role models; this requires that the director train his ear to pick up the fashion in which jargon is used in a given environment.

Voice-over versus On-Camera Demands

As a spokesperson, the talent may have to record on-camera and voice-over segments in the same production day. This is not the ideal case, but sometimes the talent's availability or the limits of the budget dictate such an approach. Often there can be problems matching on-camera delivery with voice-over readings.

A good approach is to narrate all voice-over at the start of a production day, making an audiocassette check copy while the talent is fresh. Then, throughout the production, the director can ensure that the style of delivery of on-camera segments remains consistent with the adjoining voice-overs in the script. Actors tend to get "up" for on-camera segments, no matter how late in the day, but often lose energy when they are asked to narrate late in the day.

If the voice-over must be scheduled at the end of a production day, the director should ensure that adequate time is allowed so that the session is not rushed. Further, he should give the talent a break prior to beginning the voice session. Using playback of the already recorded on-camera segments can also help to make sure that a "match" between the two is achieved.

CONCLUSION

In summary, the director's job is to enhance a production by placing the best performance on the screen and on the soundtrack. The best way to do this is to fully understand the script and the audience as well as the talent, and then work as the "advocate" for the talent to create the best possible working conditions that will allow the talent to maximize the skills for which they were cast in the first place.

10 Directing the Crew

In most production cycles, a director spends more time preparing for the days of production than he or she spends actually directing. Script review and casting sessions are combined with blocking decisions based on location scouts' reports or set designs. Rehearsals and equipment rental decisions, client meetings and script breakouts are geared to making the best use of the time when cast and crew are assembled and the program must be put on tape in a limited number of hours. This is the time when the director is under the most pressure, when decisions are the most final and have the most impact on the look of the finished production. This is the time when the director can benefit most from a supportive team—his production crew.

The ideal crew serves as an extension of the director's viewpoint while bringing additional skills to the production. The best working relationship between director and crew is one in which the director can easily delegate tasks to appropriate crew members, knowing that their performance of those tasks will be very close to what he would do himself. This allows the director to focus on specific tasks as they come up, without feeling that he is losing control of the production.

SELECTING CREW MEMBERS

The key to creating a good working relationship with the crew is to select the right crew members prior to the production day. The director who wants to maximize his own creative potential in a production will give considerable thought to choosing the right person for each task. This can be more complex than simply choosing the most skilled camera operator or the most experienced sound recordist.

The director should, of course, choose people he can rely on to perform the technical aspects of their craft well. If he requires specialty makeup, he should ensure that the makeup artist can create the desired effect. If a complex camera mount is required, his team should include a key grip or camera operator with experience with more than simple tripod and dolly setups. However, some of the less technical aspects of crew composition should be considered as well.

Crew Personalities

On any production day there is a complex interaction of personalities as well as physical actions. The right mix of personalities can greatly enhance the smooth interplay of client, cast and crew. The director may wish to think of this as directing a "play within a play"—trying to achieve as smooth a flow behind the camera as appears on tape in the finished program. This can guide him in choosing the best personalities for the production.

If, for example, the production involves taping an executive of a conservative corporation, the director may want a crew that can act in a low-key manner. This will help him in his own job since the crew's demeanor will have a large impact on the executive's perception of the taping process, and that perception will affect the executive's performance on camera.

On the other hand, a director who is faced with a long shooting day without much action—a multi-camera taping of a tax conference in a hotel, for example—may find that deliberately choosing crew members for their ability to enliven the proceedings will lead to a better crew dynamic. He may select an outgoing cameraman with a large repertoire of jokes to keep the crew awake during the long, uneventful production session.

"CASTING" THE CREW

The director should select crew members with as much care as he selects the production's on-camera roles. He must balance personality factors with the speed with which a crew can execute setups. If his production design requires that a great number of setups be staged in a short amount of time, and basic execution has greater weight than refined performance, he may have to choose pure skill over personality when composing his crew list.

A Key Crew Position: The Floor Director

In some production designs, particular crew positions play a more important role than others. For example, when working with a multi-camera crew on a production that requires fine-tuning actors' performances, the director should give special thought to his floor director. This is the person who will be the director's eyes and ears in the real environment that the talent experiences in the course of the taping process.

A good floor director will know how to keep the director (who must operate from a control room) in touch with the studio. He will let the director know if actors are bothered by distractions or seem tired while off-camera. This requires a familiarity with the director's style and an awareness of expectations, so that the floor director does not appear to usurp the director's position. Only an experienced floor director can bring the right combination of initiative and tact to a situation such as this.

The floor director is particularly important when dealing with nonprofessional talent. The floor director who relays the director's suggestions will often act as translator. He or she will get the production language shorthand from the director via a headset and quickly translate that information in language that is clear to the nonprofessional. This takes patience and an awareness of the nonprofessional's state of mind. A floor director who has only worked on productions with professional actors may not have the necessary sensitivity (or ability to translate) that is required. (See also Chapter 8, "Directing Nonprofessional Talent.")

The Director-Crew Match

Each director will have his or her own style of relating to a crew, depending on previous experience and particular production interest, but most directors will have one thing in common. Over time, they will develop relationships with particular crew members. One gaffer's personality and lighting technique may suit a particular director's style. A director may favor the naturalistic makeup that one makeup artist creates over the high-fashion approach of another artist. In any team effort, some people work better with certain others, share a more similar viewpoint and get more work done more quickly when paired with the right persons. Ultimately, a more creative program can be produced when the right director-crew match is made.

Crew Interaction

Certain positions rely heavily on interaction with people in other positions. Gaffers must be able to work well with best boys (assistant gaffers) and key grips. Sound recordists and sound mixers have to be able to work well together. (In some crew designs, specific crew positions are crew chief positions. A gaffer may hire his own support crew. On a complex sound job, the first sound position will hire assistants in the other positions, and so on.)

I make certain that I book the key crew positions first, so that even if I am booking all the crew positions myself, I will be able to consult with key crew members to ensure that I create a crew that works well. I will try to provide the persons responsible for major portions of the production with the support team of their choice. Typically, this will mean that they suggest or choose the people with whom they have experience working together.

Common Experience: Common Language

In general, the more often crew members work together, the more effective they are as a crew. The reasons for this are quite simple. A common production language develops as people work together. As they encounter a particular production situation on a program that is similar to a past situation, they are able to draw on the prior experience to come up with an immediate solution to the current problem.

In fact, a form of private language, or shorthand, based on their common experiences, allows a crew and director to spend less time working out solutions to logistical problems and more time on creative additions to a production. By communicating quickly, the director and crew are able to maximize their energies—they are able to work from a more complete picture of the director's concept from the very outset of the production.

PRE-PRODUCTION DIALOGUE

Whenever I have a production ready to book, I like to create a crew list with my first and second choices for each crew position. I will then review the list to visualize the crew interaction so that any concerns I might have regarding the crew's behavior will occur to me. This allows me to have an agenda in mind when I call each prospective crew member. During the initial call cycle in which I book the crew members, I am able to give a capsule description of the shoot. This is my opportunity to let each crew member know what my specific expectations and areas of concern are.

For example, on a production that involves children, I may spend time discussing the schedule with the lighting crew. I will point out that our lighting design will have to allow for quick setups so that we don't lose the children's energies while they wait for re-lights. In the discussion, the lighting crew will get a good sense of my concerns so that they will be able to adjust their priorities accordingly during the shoot. Without this discussion, their natural tendency would be to spend more time adjusting small details than I might be willing to allow.

Eliciting Crew Input

This initial conversation is also the opportunity for me to ask for the crew's input on the production. I have found that the crew members will be much more energetic and helpful if they are asked to help solve production problems and provide creative input before the production day itself. For example, on a program that involved extended dialogue in a moving vehicle, I spent some time discussing options for the audio design with the sound recordist. I outlined how I wished to tape the scene and how I wanted to edit the scene, while giving him a complete description of the interior of the vehicle.

This allowed the recordist to present me with several alternative audio designs and discuss the post-production requirements of each. All our conversations occurred several days before the production so that the recordist could arrive at the production site with a plan with which he was comfortable, and one whose implications I was fully aware of. Without a discussion such as this, though, I would have lost a great deal of time on location while verbally working through several options before arriving at one that worked for both of us. The time was better spent dealing with concerns specific to the location setup that could not have been anticipated prior to our arrival on site.

Pre-production discussions with the crew allow the crew members to make recommendations which might not be possible otherwise. In the above scenario, the sound recordist might have recommended the use of some new microphone that would require a rental, but would make for a better recording. Without the initial discussion, we would not have had the opportunity to secure the appropriate mike.

This is of particular importance because the crew members are hired for their individual areas of expertise. They are able to maintain a deeper degree of specialization than the director. As a result, they are more aware of the most recent innovations in their specialties and can bring this expertise to the production. But, this will be used to its fullest advantage by the director who includes them in the pre-production phase. He will also get the most creative involvement from them if they are aware that they are able to control some of the production design.

Too often, a crew member encounters an otherwise avoidable problem that stemmed from the director's ignorance of his area of expertise. The specialist who could have anticipated the problem wasn't involved in a pre-production discussion, and spent much of the actual production day thinking, "If only I'd known about the background noise (or the lighting problem, etc.) I could have brought a directional microphone (or the right filter package)." This sort of problem lowers more than the sound quality; it often lowers crew morale and decreases attention to other production details.

By spending adequate time in pre-production discussions with the crew, the director sets a standard of preparedness and creates a tone of professionalism. This will provide a good footing for the entire production. The crew members will know that they are expected to ask questions at the right time and to anticipate, avoid, and, if necessary, deal with problems quickly. They will also realize that the director values their input and that they are being given the opportunity to demonstrate their best work. Participating in the planning process gives the crew more personal involvement in the production and more of a personal commitment to creating the best product.

A Cautionary Note

The director should be careful not to overstep the locally accepted norm of pre-production involvement, however, and should recognize when he should offer to compensate the crew members for their time. If he asks a cameraman to conduct tests of various filter options specific to an upcoming project, he should expect to be billed for the cameraman's time. These expectations will vary with the director's relationship with the crew members, their personal interest and the demands of the pre-production task, as well as with the accepted work conventions in the area.

Pre-Production Communication

In all pre-production conferences, the director should leave the crew members with a clear sense of his priorities, letting them know where he would like them to spend the most time and attention during the production itself. This will produce a "directed" crew even before the production day. The basic production logistics should be covered as well: where the location is, and how to get there; what the call time is, and what the first task is (in case the director arrives late).

I have also found it useful to describe the production day so the crew members can plan their personal lives around it. If I expect an overtime situation, I will tell the crew so they can avoid the disappointment that comes with missed dinner engagements. I will indicate temperature extremes—such as on the top of windy buildings or in food lockers—so the crew can dress appropriately. I will warn the crew if a client who is difficult to work with or who must be handled with care will be present. By the same token, if I know that the production will be relatively relaxed, I will let the crew know this so that they can anticipate an enjoyable day.

Crew Wardrobe and Its Impact

I often specify a type of dress to the crew members, so they can make a good impression on a client, or lessen the anxieties of a nonprofessional who is to be taped. The client, interviewee or corporate CEO nervous about appearing on tape is not likely to be relaxed if the crew seems to come from a world with a different sensibility. All my work to gain his or her trust might be undone if a crew member wears wild clothing. Crew members have always cooperated when I have asked them to dress appropriately for a specific production, even when this has extended to wearing tuxedos to tape documentary-style at a private black-tie function.

Providing a Context

I have found that visual information pays off in a quicker understanding of what I intend as an overall effect. If I have sketches of a set, I will send them to the lighting director. If I have drawings or photographs of a location, or storyboards for a production, I will send copies of them to the key personnel so that they can have a general sense of the production before they arrive on location or at the studio.

This lets them work from a comprehensive creative context as we begin our first setup. The suggestions they make stem from a larger viewpoint. They have had time to consider the entire production and develop a sense of an overall context within which the first setup fits. Any new suggestion or creative addition that a crew member makes is then more thoughtful, not an isolated reaction to an initial scene or setup.

I have also found that including key crew members in the staging process of a production indicates to them that I respect their opinions and want them to feel free to add their experience to the show, rather than simply execute my suggestions. However, you cannot include them in this process without giving them the larger view. In the absence of sufficient information, even the most creative crew person can offer little other than technical suggestions regarding the best way to execute the director's vision.

Helping the Crew "Visualize"

When working on a production that involves multiple setups, I have found it useful to involve the lighting director in a location walk-through or a production design meeting. This provides an opportunity to suggest the best sequence of staging scene setups from a lighting standpoint. Potential problems can be pointed out and natural lighting sources can be considered. If budget allows, I may include set and prop persons and stylists in this phase.

This creates a forum for previewing the entire production in terms of a visual design. The final screen effect becomes real and we are able to deal with staging in isolation from performance and content concerns. Most often this results in a program with a consistent, and often unique, "look" rather than one that looks like many other programs—designed moment to moment and scene to scene during the production itself.

CREW MEETINGS DURING PRODUCTION

No matter how rushed the beginning phase of a production is, a director who wants to get the most performance from his or her crew should initiate a meeting at the start of every production day. This may mean a single meeting with all members of the crew or it may mean a series of smaller meetings. The lighting and grip crew may need one rundown, styling and makeup another, and teleprompter and talent yet another.

Initial crew meetings are the forum for the director to map out the day's requirements so that crew members can pace themselves accordingly. If the crew knows which tasks are critical at the outset of the production, they will be able to focus on them at the appropriate moments and not spend too much time on smaller details. The initial meeting also allows the crew to "warm up" to each other and to the director.

Ranking Tasks

The initial meeting is the director's opportunity to find out who is going to be a good listener and who is not—even to find out who has had a long shoot the night before and who is well rested. Knowing this may affect to whom he assigns a particular task.

I may approach a production with the idea that the most experienced camera operator will get the most difficult position; but if I find that he has worked much of the prior night on an overtime production, I may assign that position to another camera operator who is fresher. The reason for this choice would be known only to me, however, because I would simply tell each camera operator what camera I would like him to handle before I described the task for each position. This way, I do not get into the position of passing obvious judgment on their respective abilities in front of the other crew members.

THE IDEAL PRODUCTION PLAN

It has been helpful for me to look at the typical crew's day in terms of the flow of attention and energy and plan some production tasks accordingly. Obviously, some logistics, such as the position of the sun, aren't controllable, but it is useful to consider the typical flow of a crew's energy and attention when scheduling those production elements that can be staged at any time of the day (see Figure 10.1).

Figure 10.1: Typical Flow of Crew Energy

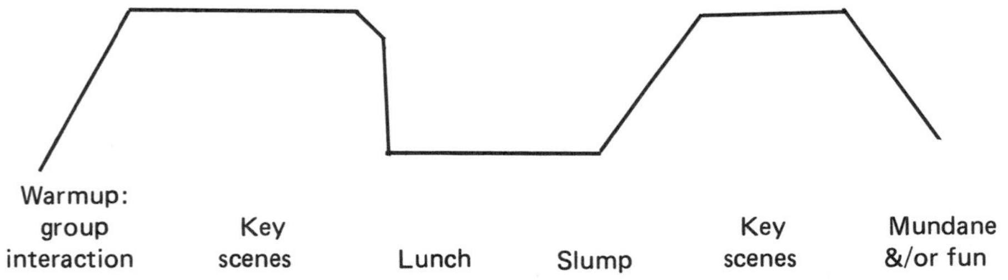

Warmup:
group   Key   Key   Mundane
interaction   scenes   Lunch   Slump   scenes   &/or fun

The Warmup Period

In the first hour or so of production the crew members not only finish waking up and become tuned to their surroundings, they also create a pattern of communication with each other and with me. If we're on location, they must unload equipment and locate power. Technical details such as camera warmup and registration, monitoring hookups, sound checks and tape checks must be done. Talent must be situated and makeup applied. Set elements must be placed and arranged, and windows must be dressed. These are just the technical adjustments. As a group of people, the crew is making its own human adjustments during the initial phase of a production day.

Crew Adjustments

We develop a more refined language regarding the production as the day unfolds, because we are, in many ways, seeing all the elements in one place for the first time. We are also adjusting our sense of timing. This is particularly true for live multi-camera productions; but even on single-camera film-style productions, the camera operator and the talent adjust to each other's timing and both adjust to the director's timing cues as the day goes on. These adjustments may be very subtle but their cumulative result can be a more believable or more dramatic or smoother interaction as the team begins to "click."

Since this period of adjustment is predictable, I try to avoid scheduling critical scenes during this time. Instead, I will make our first setups or scenes the "middle of the road" scenes that involve basic characterization or standard camera moves. This allows the crew to begin to reach a working pace without attempting a particularly difficult task "right out of the gate." It also lets the talent warm up as well.

Initial tasks should offer enough challenge to create the working spirit that pulls a crew together, however. If the initial tasks are so mundane as to offer no challenge, it may be difficult to get the crew "up to speed." I like to think of this phase as the opportunity to build the basic interaction patterns for the cast and crew—and often the client, as well.

Peak Performance Opportunities

By mid-morning on most productions, everyone's adrenaline is flowing and each person has met the other members of the crew and has developed a working relationship. Basic timing adjustments and overlaps in perceived areas of responsibility have been worked out. The crew has begun to act as an efficient team that executes setups and takes with a minimum of false steps. When the director calls for a change, several people spring into action without falling over each other. In a word, the production team hits its stride.

This is the time that I have found best for scheduling the key scenes which require the most concentration and attention to detail. The "workings" of the crew, the interpersonal logistics and the technical requirements have jelled. Everyone on the crew is now able to focus on what is actually being created on tape, and not so much on the details of how it is being created. There are fewer distractions because the crew members do not have to interact verbally on the small details. Each crew member understands the level and style expected in the production and begins to attend to details without being told by someone else.

The more dramatic scenes, which involve complex moves by the camera crew and subtle lighting details by the lighting crew, are done most effectively when this point has been reached. For example, a sound boom operator may have to work a shot between light stands while camera dollies and actors hit their marks while delivering dialogue with the right timing and intonation. The complex interaction involved is best when the crew has reached a peak level of performance.

The more subtle characterizations by the talent should be left to this period because the director and actors will have fewer distractions and will be able to notice the details as the scene is being played. When distractions abound, more tape playback is required and this results in a slower rate of production and less effective interaction between director and talent. Decisions will be made more quickly during this phase.

The director who has scheduling control should take advantage of this predictable phenomenon and create opportunities for doing large amounts of good work in this time frame. The simpler, more mundane tasks can be scheduled earlier or later in the production day.

Pre-Lunch Planning

On many productions, the lunch break is a time for cast and crew to relax and interact with the client on a social level. I have found that it can be very helpful to create an "opportunity for accomplishment" before a lunch break by choosing to end the morning session with a scene that is difficult. This can be very satisfying for the cast and crew.

It gives them, and the client, a solid sense of accomplishment to take to the lunch break. It will foster camaraderie and a pleasant tone during the more relaxed time spent over lunch. The pre-lunch scene might also be chosen as one that is fun to do. It may involve a unique production concept or humorous interplay.

Creating a good interaction over the lunch period can help foster a more spirited sense of teamwork throughout the rest of the production day. By the same token, scheduling a mundane or depressing scene prior to the break can have the opposite effect.

The Post-Lunch Slump

Everyone involved in a production has, at some time, experienced a common post-lunch pattern: a simple task suddenly becomes difficult, crew members or actors suddenly seem slower than they did just before lunch. Everyone expects to regain his or her former momentum immediately, yet, for a short period, nothing seems to go right. It is common wisdom that many accidents occur right after a lunch break. (There is a simple physiological reason for this pattern. On any production, the crew expends a great deal of energy moving themselves and heavy objects. They use up a lot of calories and eat a hearty lunch. After lunch, their brains lack blood and oxygen while their blood goes to their stomachs to aid in digestion.)

It often takes a half-hour before the crew can achieve the same unity of performance as earlier. This is not the time to send someone up a tall ladder to adjust a heavy light or to expect perfect execution of difficult lighting and camera cues. It is a good time to plan some of the simpler tasks or to plan a setup that requires a lot of preparation time. The experienced director will realize that this is a normal occurrence and will not become unduly upset. The crew will regain its ability to perform very quickly once their meals are digested. To a lesser degree, after any major break, the director should expect the crew to require an adjustment period before it regains the cohesion and solid interaction that was built up prior to the break.

The Later-Day Period

As a production wears on over a long day, the crew rarely has a true sense of how much more there is to be done, no matter how much involvement it has had in the pre-planning process. The director may have added and/or dropped shots that the crew is not aware of; the last setups may be simpler or more complex than they can guess.

The director, on the other hand, should have a clear sense of the amount of work required to finish the day and he is usually quite motivated by that goal. Also, the director is in a more active role throughout the day—interacting with talent and client as well as crew. He has a ready source of adrenaline in his motivation to get "his" program done.

This often results in an imbalance in energy levels toward the end of a production day. Many enthusiastic directors find (to their apparent amazement) that their crew does not respond with the same enthusiasm throughout the day. Consequently, it is important for the director to create a realistic schedule for the later hours of any production. Otherwise, he will begin to run late, find that the crew works less efficiently as they get tired and will run even later on the succeeding setups. This can lead to a downward spiral that can require a choice between heavy overtime or the director's acceptance of poor-quality performances.

Besides scheduling realistically, the director should plan on more frequent breaks for the crew as the day wears on. The crew, not as goal-aware as the director who needs the tape for his edit, is less able to ignore fatigue in the later part of the day. Since the business of video production involves hot lights, physical labor and frequent conversations, an active crew is also prone to dehydration and should be provided with water and other liquids besides coffee. I have found that, even when we are close to the finish of a production, it often makes sense to take a strategic break. In the long run, we finish sooner with better results than when we continue to work steadily without breaks.

Planning the Wrap Process

In the latter part of a production schedule, the director should look to ways in which he might facilitate the "wrap." Too often, productions end with a large amount of equipment to be packed by a very tired crew. The longer the period of time spent in any one location, the longer the wrap will take, because people tend to set aside items without packing them in the rush of production and there is a cumulative effect of this inadvertent procrastination over time.

I once worked at one location for four days straight with eighteen actors and fifty-two setups. The full "wrap-out" of the location took almost six crew hours since the schedule had been a "rush schedule" and very little "wrap-as-you-go" was possible. (In an instance such as this, it sometimes makes economic and logistic sense to hire additional help for the wrap process, or to schedule it on a separate half-day with a fresh crew.)

One way to facilitate the wrap process is to schedule silent, or "MOS" scenes, at the end of a day so that gear can be packed up and the noise of that process will not interfere with the recording of scenes. MOS refers to scenes that are shot silent. The term derives from a German director in Hollywood years ago who used the term "mit out sound" (MOS) to differentiate between those scenes requiring clean audio simultaneously recorded with the picture versus scenes during which crew and cast could make noise without worrying about its being recorded. This distinction is an important one since a director and crew can give precise verbal directions to each other and the talent during filming or taping an MOS scene, which often simplifies production logistics.

For example, reaction shots can often be done at the end of the day, freeing the audio crew and allowing crew members to make noise while they work. Smaller-scale scenes which require less equipment can be scheduled so that unneeded equipment can be struck. Simple graphic shots can often be done with a minimum of lighting and grip equipment, as well as with less crew, so that the bulk of the wrap can proceed during the taping process. The director who has used a cameraman/gaffer may be able to assume camera duties so that the gaffer can supervise the lighting crew's wrap.

I have also found it rewarding to schedule any offbeat or humorous elements for times when the crew needs a lift. (This might mean letting an actor play out a scene in the outrageous fashion he has been threatening to use all day, or actually planning to do any called-for scenes with a humorous angle.) This will go a long way toward lightening the burdens of coiling cables, packing thermodyne cases and hauling sandbags.

DIRECTOR-CREW PSYCHOLOGY

A director will get more creative effort from his crew if he keeps several important guidelines in mind.

Developing a Directorial Style

As the director sets about finding his particular style of communicating he discovers, sometimes through trial and error, which crew members work best with him. There will be times when he knows he has the wrong persons on the crew and times when he will miscommunicate. But if the director is able to work through this period to develop a leadership style that works, he will eventually have the satisfaction of creating a team that is an extension of his own creative impulse. This will allow him to craft more refined productions that are consistent with his own vision but have additional nuances that talented crew members with initiative bring to them. This, in effect, amplifies his creative efforts.

The video director who works into this role would benefit from the traditional studio-apprentice approach in which a director first learns his craft by observing other directors' working styles. However, since many video directors find themselves in a responsible role without benefit of an apprenticeship, they must rely on observation, basic human psychology, and their own abilities to learn quickly from their good and bad decisions.

Background Factor

This newer tradition also means that there are many self-taught directors who slightly redefine the director role to suit their own particular area of interest or expertise. My own style derives from an academic background in philosophy and psychology, which allows me to handle the personal interaction aspect of crew direction from an analytical standpoint. My experience as a cameraman lets me be very technical in describing visual effects, but I have no training in set design or styling so I rely more heavily on stylists and designers to give the main direction in visual design. Since I have edited my own programs, I will take an active interest in continuity considerations.

By contrast, another director who comes from an artistic background will deal more actively with scenic or set design elements, engaging in lengthy conversations with stylists and prop persons to assure that those details are to his or her liking. That director may pay very little attention to continuity from shot to shot, relying on a script supervisor or continuity person to provide that check. Yet another director may focus more on the sound texture of a program, working very closely with the sound recordist, audio engineer, or sound mixer. A director who is not strong in the area of crew direction may rely more heavily on a production manager or producer to assist in coordinating the crew.

The range of directorial styles is practically as diverse as the number of directors creating programs. Each has a different way of relating to the set of human beings who comprise the crew and will interact differently with them. Most will probably follow some of the same basic guidelines to get their final results, however.

The Delegation Dilemma

Working with crew members to give them an overview and allow them to participate in the overall program design lets the director extend his creative capabilities. In order to involve crew members on this creative level, however, the director must do two things. He must create his own vision first, so that he can communicate a consistent picture of the production as a whole. At the same time, he must be able to give up his direct control of small details. The latter often presents a problem for a director who is not accustomed to working with larger crews.

The Transition Period

The director who is "growing into" larger productions has a history of doing most crew tasks himself. He has built the habit of doing rather than telling. He has done camera framing and slight lighting adjustments over the years without even being conscious of doing them. This means that he hasn't built a vocabulary for communicating his wishes to others. As a result, it will initially be difficult for him to translate his wishes into words.

There will be a period of some frustration as he adjusts from the more solitary role to the crew directorial role, because he will have to be more conscious of his role and think before he acts. Initially, he will feel that activities actually take longer and will have to resist the impulse to perform them himself. However, by taking action himself he will be sending a message to his crew members that he finds them inadequate. (The crew that doesn't feel respected by the director will not, in turn, perform in a fashion that gives the director confidence in its members abilities.)

Helping Crew to Look Good

All people in the video production business value their self-image. The entire business is based on creating good images, and everyone on a crew works to create the best image before the camera.

Intelligent, talented crew members will respond very well to a supportive atmosphere in which individual achievements are recognized by their peers. The director of any production can set the tone for this atmosphere by the way in which he gives direction.

Give Crew Members the Information They Need

The key to this is to provide enough information at the outset to let the crew members execute directions with a minimum of correction by the director. This is simply a form of making sure you, as the director, communicate well. If you give clear direction, the crew members will respond quickly and will get their own positive reinforcement from their peers' mutual recognition that they are looking good as a team. (This has other, less tangible, benefits as well.)

Provide shot-sheets for camera operators on a multi-camera crew. Give makeup and wardrobe personnel information about set colors as they work with the talent. Let engineering staff know what out-of-the-ordinary technical requirements you might have, and when they might occur. Do all the *pre-deciding* and *pre-planning* that will let the crew stage work efficiently.

For productions with multiple film-style setups throughout the day, it is useful to create memos that describe the setups. Then, individual crew members can use the time between takes to prepare the elements that are involved in the succeeding setups and strike the elements no longer needed in the present setup. This lets them move quickly from setup to setup. This also allows the director to spend less time and mental energy on explanations and more time on the creative aspects of the production, refining setups as they come together.

Give Crew Members the Time They Need

Besides giving crew members the information needed to do the job, the director should provide them with realistic schedules. He should learn enough about each crew member's job to make realistic allocations of and reasonable demands on his or her time. He must also learn what each crew member means when he or she says that a particular task "won't take long," or will "just take a minute." This lets him plan the day better, as well as make appropriate adjustments during the production.

A director can make his engineering staff look better, for example, by giving them as much time as possible to set up cameras. This can be done simply by learning to rehearse without cameras until the last possible minute. It also requires that the director practice a discipline of rehearsing in stages, saving some adjustments for later steps.

Act as Traffic Cop

Keeping set areas clear for lighting crews to work quickly, directing clients out of the crew and talent's way, calling for quiet when audio needs to make a check—and generally directing the setup process as well as the scenes themselves—will pay off with more efficient performances by the crew.

By providing working time and space for all personnel, the director lets them show their best work to their immediate audience—the client and any onlookers. The crew gains respect and support. The crews I have worked with have appreciated the opportunity to be seen doing good work.

Rehearse in Stages

Blocking action in a rough fashion with the camera crew before running actors through full rehearsals allows the crew to get the basics down first, without tiring the actors. Rehearsing a compound camera move for framing stages, for example, before combining it with lighting cues and camera moves, will allow the crew to find the right solutions as they go, so that the first full rehearsal looks as good as possible. This will require fewer refining rehearsals, using all the crew and talent, and will keep everyone fresher through the course of the day.

Admit Errors

There are some directors who, in their own eyes, never make mistakes. They are not usually well thought of by their crews, because they typically blame crew members rather than themselves for all mistakes. Their motivation is often to look good in front of their clients. However, even the least sophisticated client quickly learns to read crew reactions to a pompous director and deciphers the underlying reality.

If only for self-protection, a wise director will admit his own errors rather than blame them on the crew. Crew members who are falsely blamed for the director's problems will invariably have many opportunities to return the "favor." (The live director who constantly blames a technical director for his own miscalls soon finds that the technical director is not likely to "save" him from his own mistakes, such as calling for a nonexistent camera or telling him to "take 3" when he is already on "3.")

Provide Consistent Timing and Terminology

Every crew develops habits based on its interaction with the director. The more consistent a director is in his camera calls, countdowns and action cues, the more quickly the crew responds with the right action, at the right time. Consistency is everything when working with a crew, since every production situation is slightly different and the crew is always "learning on the job." Consistency allows you to set a professional tone that enables people to work effectively together. Changing terminology in the middle of a production can cause major confusion on the crew's part.

I once acquired the habit of using "wuff" to mean "stop zooming" (or panning, etc.) when directing multi-camera live productions with the same crew on a regular basis. A new camera operator joined the crew one week and in the course of the production I asked him to zoom in while his camera was on line. As the framing became what I wanted, I called "wuff" over the headsets. But the camera kept zooming in. After my third "wuff" with no response (he kept zooming right into an extreme extreme-closeup) I finally called "stop" much to the camera operator's relief. He'd had no idea what the term "wuff" meant.

Share the Clients' Appreciation

On most productions, I am the only person on the production crew who sees the program through to the final edit and delivery to the client. The crew members usually do their jobs and move on to another production. I have learned from other producers and directors the value of staying in touch with a crew after the program wraps. One way is to call every crew member after a production is completed and let them know briefly that it was well received and that both the client and I appreciated their efforts.

I have found that most crew members are pleased to hear the final results of a show they have worked on, and to receive awards, kudos, or simple thanks. The goodwill that results from this courtesy makes it well worth the effort.

The Larger Framework

I have found it most useful to approach any crew as a group of professionals in the larger framework of working within a given production community, rather than as a collection of talents only brought together for the one production at hand. Showing respect for a crew's abilities and making an effort to share compliments with them immediately pays off in better working relationships over time. Giving them the recognition they deserve creates a good working atmosphere that carries over from one production to another.

Cultivating the skills of newer crew members allows a director to develop an increasing talent pool with "bench strength"—enough people with varied skills who know his directing style. He will then be able to create a supportive crew in a short time to respond to production opportunities. This is what ultimately gives a director the confidence to attempt new production techniques. That confidence, in large part, is the basis of a director's creativity.

11 Directing the Edit

INTRODUCTION

The most exciting point in any video production is the editing process. Here, the director sees many heretofore disparate elements come together into a cohesive whole in a very brief time. Audio is matched to video, narration and sound tracks are heard at the same time that pictures are seen. Elements shot out of sequence are placed in their final relation to each other. Timing between cuts is established and the program's pacing is developed.

The edit process provides a great learning experience for any director. During the editing phase he finally sees how well his previous visualization works in its final realization. The nature of video editing allows this to occur in a very brief span of time. There is no need to send film to a lab, cut a workprint and conform other elements in multiple stages. For the most part, the video edit combines all elements quickly, with the opportunity to make changes "on the spot."

Early in my career I worked primarily as an editor. I spent many hours editing for clients at a post-production facility. Most of these were producer-directors and this gave me the opportunity to watch their various methods of directing the edit session. From this, and my own experience as a director-editor, I have seen the value of developing a systematic approach to the edit process. This approach involves the preparation and organization of scripts, script notes and other production elements to make the most efficient use of the time spent during the edit process. This approach allows directors to enhance the production value of the programs they direct and to maximize their clients' post-production dollars. This is because the adage "time is money" is most true in the post-production phase of a project.

AN APPROACH TO EDITING

Developing a systematic approach is actually very simple. It involves organizing the appropriate elements involved in the edit so that each editing step can be performed as quickly as possible. This gives the director and the editor the opportunity to spend more time experimenting with valid edit options. At the same time, predefining a limited number of workable edit options will allow the director and editor to devote their energies to only the most rewarding of those options. The net result is that the director's and editor's energies will have the maximum impact on the final, on-screen product: the finished program.

The Editing Flow

There is predictable "flow" to each editing process. To get the most value out of a program, the organization for each step in the editing process should be guided by that flow. For example, during any extended edit session for a program, an editor develops a particular rhythm. Loading reels, cuing shots, previewing selected takes and trimming edits are the tasks involved in the basic cutting of a program. These are quite different sorts of activities than composing graphics on a character generator, equalizing varying audio segments, formatting effects, etc. Shifting gears and changing tasks frequently slows down the editor.

This can also create inconsistencies in the final product as well. Every time an editor has to switch his attention from color matching to sound mixing, for example, he has to reacquaint himself with the overall flow of either the audio or video portion of a program. This makes it more likely that he will mismatch sound levels, video levels or colors. The director will create a better program from a technical point of view if he is able to consolidate "like tasks" for the editor and create a smoother flow in the post-production phase. (This holds true whether the director is his own editor or he directs an editor other than himself. We will assume the latter for our purposes; but note that these principles hold equally true for either situation.)

Preparing for the Edit

The most effective way to facilitate a smooth editing process is to build a note-keeping system at the start of the production that will allow the director to keep track of his ideas for future audio and video elements. It should also act as a way to remember his judgments from the set or location during the production. For example, Take 3 of a given scene may have been acceptable in terms of talent performance, but the camera move may have been marginal in the director's judgment, which led him to call for a Take 4. In Take 4 the camera move may have been perfect, but the talent performance may not have been as polished as in Take 3.

With a good notekeeping system, the director can quickly reconstruct the decisions that were made on location without reviewing takes or searching through paperwork. He will be able to review only those takes that are possibly useful, as determined during the shoot, and will be able to focus on the questionable aspects of each take during his first review of each "select take."

Daily Updating of a Working Script

There are several ways in which this task can be performed. If time and/or crew size allows, a complete linear list of all shots should be kept as "camera notes," and the possible selections should be transferred to a working script at the end of the shooting day. (Traditionally, this task is performed by a script supervisor or continuity person. On a controlled, moderately paced production, the director may be able to do this himself.)

This will allow the director and editor to refer to the "selected takes" while working from just one completely annotated script. They will not have to spend time reviewing rejected takes and remaking the decisions that were made during the production.

Other Updating Techniques

If the production schedule does not allow this approach, then an alternate method can be chosen. The director may ask the VTR operator to keep a linear list of all takes (the camera notes, or "shot log"), noting his directorial comments as they progress. The director simply has to call out which takes, in his judgment, are potentially usable. Then, once the director is through with a scene, he can ask the VTR operator to give him the time code for those takes. The director then writes those take numbers and their corresponding time codes on his working script. This will later serve as the master script for the editing process.

Another way to accomplish this is for the director to call out at the end of each take whether the take was usable or not. Then he should allow time at the end of each setup, or at convenient break points in the production, for the VTR operator to transfer the selected takes and their corresponding time codes onto the director's working script. This can be done while the director is occupied with tasks that do not require him to use his own script.

Yet another technique is for the director to ask his tape operator, who is keeping the camera notes, to annotate each take with the director's comments. Then, prior to the edit, the director transfers time code and comments regarding selected takes to the master script. This can help the director refresh his memory regarding the takes and their content prior to the edit without having to review the takes themselves. (On a production with about thirty-five planned edits for a seven-minute

Figure 11.1: Samples of Camera Notes vs. Master Scripts

Camera Notes

SCENE	TAKE	DESCRIPTION	COMMENTS
8	1	Host emerges from lobby of building and introduces program segment on camera.	NG (No Good): INC DIAL INComplete DIALogue Also known as a "broken" or "busted" take.
8	2	Same action.	NG—Airplane noise.
8	3	Same action.	Good.
8	4	Host continues to walk during 2nd half of lines.	NG—Camera mistake.
8	5	Same action.	Good.

CORRESPONDING SCRIPT WITH TRANSFERRED NOTES

SCENE 8

Spokesperson appears on camera as he steps out of corporate headquarters.
T3-[2:11:42—8-3]
[W/O walk]
T5-[2:13:50—8-5]
[with walk]

YOU ALL KNOW THAT OUR COMPANY HAS A LONG HISTORY IN THE FASHION INDUSTRY. BUT YOU MAY NOT KNOW THAT WE WERE ONE OF THE FIRST COMPANIES TO ADAPT THE NEW MANUFACTURING METHODS PIONEERED IN THE AUTO INDUSTRY. THIS ALLOWED US TO INCREASE OUR PRODUCTIVITY QUICKLY AND TO RAPIDLY BECOME AN INDUSTRY LEADER. LET'S TAKE A LOOK AT THE IMPACT THAT THAT EARLY DECISION HAD ON THE FIRST EMPLOYEES AND THEIR FAMILIES.

marketing program, I spent an additional two hours transferring the notes. But it enabled me to become "re-familiarized" with the material prior to the edit and greatly increased the pace of the edit itself.)

Advantages of Annotated Master Scripts

A quick glance at the script in Figure 11.1 lets the director and editor know that only Takes 3 and 5 of Scene 8 (8-3 and 8-5) are usable and worth reviewing. Further, they know exactly where these scenes are located by time code and can use the "go-to" or cuing function of their edit system to quickly find the start of each take. In addition, they can look at each take, knowing ahead of time the differences between them, and can make a decision more quickly because of this fact. Having all elements, including dialogue, before them as they do this lets them focus their attention and progress quickly through their review.

If this program were 20 minutes long, with 60 or more individual scenes, the amount of paperwork that would comprise the camera notes would be considerable—possibly 50 or more pages. To look for any one scene's notes (when the production is shot out of sequence, as most are) would require someone to look through the notes, which are organized according to the shooting sequence.

This often requires people to remember when certain scenes were shot. This is not always an easy task after the production has wrapped. The reasons for certain sequencing of scenes are not that apparent after the fact. They also would have to decipher notes that were written in the haste of production. All of this takes time away from the actual editing and review of edits.

It is much easier, more efficient and more enjoyable to work with a script that has all the necessary information and *only* the necessary information. This also avoids the temptation to simply look at all the takes off-tape prior to making a decision—a common method used by persons without an organizational system that often leads to a great waste of time and dilution of energy. The director and the editor spend an inordinate amount of time reviewing the elements of a program, becoming distracted by the good and bad points of unusable takes in the process.

THE POST-PRODUCTION ENVIRONMENT

All the above makes logical sense, but there is even more to the situation. Let's look more closely at the editing environment itself. In the editing process, editor and/or director are in a semi-darkened room so that they can see the details on the video screen. Their eyes gradually adjust to the room light level and they become very focused on details within the screen frame. (That is the intent of the design efforts spent on post-production environments—to enhance this focus of attention.) The screen and its contents become the dominant reality for the persons involved in the edit.

Editor and director try to shut out the flow of the day-to-day world so that they can pay particular attention to the pacing of the program they are creating. This is key, since a program can be radically different in its impact on an audience as its timing is changed from scene to scene. The magic of editing is that it can enhance scenes. The difference of a few frames, tenths to thirtieths of seconds, can make a great difference in whether a given edit seems natural or not.

As the editor and director become more focused on the program in the making, they become attuned to the program's internal timing. This allows them to know immediately, upon previewing an edit, whether its timing is correct or not. Every preview consumes many seconds of time outside the content that is being edited: that is, the tape machines must cue up to a designated point, pause, cue back for their pre-roll, pause, roll forward to preview the edit, then pause again, awaiting a decision from the editor/operator. This means that every time the director and editor must look again at a preview option they must consume a designated amount of time. Consequently, distractions and loss of focus of attention can cost a great deal more time than the distraction itself.

The Impact of Distractions

Searching through paperwork is a great distraction in itself. The director's attention shifts from the video screen to a printed page. His eyes adjust to a lamp's illumination of printed information. His sense of timing shifts from the imposed reality of the program's pacing to the "out-of-time" reality of reading a printed page. The process of looking for a particular scene's notes in print takes away from the focused attention that is required for good editing.

Something else occurs as well. The editing process is one of "instant gratification": a decision is made regarding the timing of an edit and it is immediately laid down on the master tape, available for viewing. This creates a pattern of, and expectation for, quick results. As director and editor search for shots without finding them, they become impatient with the lack of results. If a great deal of searching is required, then they become less inclined to review shots often to look for the one best edit timing. Basically, the more they are required to search for shot information on paper and on tape, the less critical they become in their judgment of their editing and, ultimately, the quality of the program suffers.

So it is for more than simple economic reasons that edit notes should be well organized. It is also necessary to the dynamics of the process of creating the best quality program in the final edit.

THE EDITING PROCESS: SCENE REVIEW

If time allows, the director should look through all the selected takes of principal master shots or major scenes to reacquaint himself with the overall feeling of the program. This should be done prior to any editing. (As we noted above, the organizational option in which the director transfers select take notes to his master script builds in a part of this process.) The reason for this step is to get "the big picture."

Editing is, by nature, a linear process in which each shot is edited onto the end of the prior shot (or, less often, inserted into a shot). This means that decisions are made on the small scale. That is, a shot will be selected because it cuts best to another scene, due to particular movement in the frame. Or the occurrence of dialogue during action may dictate that an edit be made at a particular place.

Sometimes a series of choices made on the small scale will not be the best decisions made in light of the program as a whole. There may be dramatic reasons for selecting a take that is harder to cut well but which is actually more in keeping with the overall tone and feeling of the program. The director's review of the main action or key scenes will help guide him to make the right decisions that are in keeping with that larger view.

Making Initial Decisions

In this first step, the director may well be able to reduce the amount of material that will be reviewed in the next step. For example, he may have shot some scenes in two ways, as in our sample Scene 8 in Figure 11.1. In this case, he had decided on location that the scene could work with added motion on the part of the talent (one take includes the talent walking and one does not). The initial review process lets him compare the two takes (Take 3 and Take 5) and decide which approach he prefers, prior to performing the actual edit. This allows him to be more focused and to have a clearer sense of the final look of the scene within the program when he begins to make this specific edit.

Creating Select Reels

When working with time-coded camera original tapes and window dubs of them in the edit, there is one approach that is useful for dealing with a large amount of original material. That is for the director to copy over to a "select reel" only the material that he feels is required for the program. (This is a technique, borrowed from film editing, in which specific segments are cut out of the original footage and placed in a "work bin" or "select bin.")

Commercials lend themselves to this process since only a limited number of final scenes will be used, but a great number of takes may be produced. This technique can be useful for specific segments in longer format programs.

I have found it useful to create a select reel when I am dealing with a sequence such as a program open or a transitional montage that requires many edits in a short space. (These elements are very similar to commercials, in that they are usually tightly edited sequences with a large number of short edits derived from a lot of original material.) By copying only the "likely-to-be-used" elements I am able to reduce the amount of time necessary to find each piece in the edit process. (The loss of a video generation is not important here since we are dealing with copies of originals.)

Reducing Search Time

If we are, for example, editing a travelogue and one element is a transitional montage that shows small pieces from an upcoming segment on Japan, we might have a dozen or more reels of shots taken in Japan. The montage would be comprised of only the most poetic shots; so, as I review those reels, I would copy over to a select reel each of the shots that is particularly striking.

The result is that I end up with one reel that has those shots close together on one tape. This eliminates the many minutes of tape search time and reference to notes that would be required to locate those shots on their original reels. The montage then edits faster. And since a montage can be constructed without following an exact script, this approach allows director and editor to try many different options that they would not have time to attempt if the shots were on their original reels. (Again, they are also more focused on their task, since they can accomplish it in a shorter amount of time with fewer distractions, and fewer "gear changes.")

Listing Production Elements

As he makes an initial review of the footage that was shot, the director should create notes for additional production elements that will be needed for the final edit. Here is where he can save a great deal of otherwise expensive time. For example, in our Scene 8, a music bridge might be helpful, and a character-generated title may be needed to identify the spokesperson. These are items that can and should be prepared prior to the final edit session. By beginning his list early in the production process, the director can be assured that he will cover all the bases and not have to stop later in the edit to create those elements.

This may be much more important a task than it appears at first glance. For example, with any names that are superimposed on the screen, the director will have to verify that the spelling is correct and that the titles assigned to people are appropriate. If such a list is compiled early in the edit process, there will be time for all the appropriate phone calls and approvals. There will also be time to address any second thoughts that invariably occur to the people asked to review such a list. (I have seen many a $350-per-hour edit session halted while people spent time on the telephone dealing with just such matters. I have even had to leave "holes" in programs when people could not be reached for approvals, and this required a second edit session to finish the program.)

Consolidating Tasks

It is also more efficient to create a list of all the character-generation needed for a program at one time. It gives the director more control. When it is not done in this fashion, the director may not choose the graphics that work well for the entire program.

For example, a director may begin the final edit of a program knowing that he has to key in the names of various people and that he has to create a few simple graphics with words and figures from a character generator. These seem fairly simple and the director chooses to compose them "online" during his edit session. Since the edit session is linear (from the start of the program to the end, in sequence), the director will choose a graphic style when the first graphic or name key appears. In isolation, this may seem a good choice, but as the program edit progresses and successive graphics are called for, he may discover reasons for a different composition, type style, or color choice.

He then is faced with difficult decisions. Does he keep the style consistent with the work already done so that the show has one "look?" Does he use a new style for the new graphics so that they work better—and if so, does he go back to the earlier work and redo it so that the "look" of the program remains consistent?

It is far better to lay out all the needs at once and come up with one style that works well for all the graphics in the program. The program will benefit from consistency and each graphic will have been created while director and editor were focused on graphics creation (rather than having just changed their attention from editing to graphics production). They will have done a better job on each graphic as a result.

Economic Advantages

On an economic level there are additional benefits to creating similar production items all at once. Many production houses allow a discount for unsupervised work such as creating pages of character generation at night prior to an edit session, working from a printed page of instructions. For such a task, without time pressure, they may be able to use a junior editor or a trainee to do the work. Or, discounted rates may be available for supervised sessions in hours that aren't in high demand.

Even if the basic graphics must be slightly modified before or during the online session, the final costs can be far lower than those involved with creating such elements during a full online session. As we mentioned above, this is also a different kind of work and it tends to break the rhythm of an edit session, slowing down the editor as he switches gears.

Allocating Time

By listing these needs early, during initial tape review, the director can create a basic post-production needs list. He can then use the list as a guide that he can modify and add to as the rough edit or first cut progresses. After that session, and prior to the final edit, he can then make sure that all the necessary preparation is done.

He can also use such lists to help gauge the amount of time needed for the edit sessions. This is a key point, since it is not always a simple matter to allocate edit time appropriately. Some other post-production tasks that should be noted will suggest themselves during the next step—the rough cut.

THE ROUGH CUT

Since the edit process is one in which the participants learn from doing, it is often best to plan to make a first cut, or "rough edit." This edit of the program can be used as a guide; it allows the director and editor to look at the program as a whole in order to best judge which portions of the program are working well and which are not.

However, the director should not take shortcuts simply because he is doing a rough cut. He should make the actual timing decisions for each edit, since a poorly edited rough cut will create a false impression of the program as a whole. The basic content and timing of the program should be crafted as near to its final edit as possible.

Elements to Include

The elements that will be missing will be those that require the more expensive technology of an online edit suite: dissolves, special effects, wipes, graphics, keys, etc. I feel that a rough cut should include any library music that is planned and may or may not utilize original music, depending on the working style of the director and the composer. Since the client often signs off on the rough cut, I have found it helpful (from a client-understanding viewpoint) to include as much of a completed sound track as possible in the finished rough cut.

The rough cut process should allow director and editor to experiment with different options until they find the best approach to a given segment of the program. Time must be allocated appropriately for this process to take place. Rough edits should always be planned with a margin for experimentation and, particularly, with some allowance for the director and editor to find that some segments may not edit well as initially conceived. New edit strategies may need to be employed.

Director and editor should feel free to try those new strategies without seeing this as a reflection on their personal capabilities. The rough cut process is, after all, a chance to experiment. One factor which often gets in the way of this process is too-active involvement by a client.

Client Involvement

Clients often want to be involved early in the editing process. This is understandable, since the client has funded the program, seen many hours and dollars spent creating videotape footage, and is eager to get some sense of what the final program will look like. This is especially true if the client is a novice at the video production game and does not know what is expected of him or her.

However, it is often best not to involve a client in the early stages of an edit session. This is the time when the director should be able to judge, as candidly as possible, how the program is coming together. If a client is present during the early stages of the edit, the director will have to spend time entertaining and informing the client. This can intrude on the key process—seeing that the program is as well edited as possible.

It is best to let clients know early on in the production process that they will be presented with a rough cut at an appropriate time. Further, they should be assured that there will be plenty of opportunity for them to review the program with the director prior to the final cut.

THE OFFLINE SESSION

Even though it involves an element of instant gratification, offline editing is, essentially, a tedious process.* No matter how well organized a director is, he will have to wait many minutes for a tape to cue up to a specific point. Often, a director will be tempted to use these moments returning phone calls, calculating budgets, preparing for future productions, etc. *This* is a mistake.

Even though the time spent waiting for tapes to cue up seems unproductive, the overall edit session will even less productive if a director loses focus during this time. (When tapes are cued, and he is occupied with a phone call, he will either direct the edit with split attention, or cause the edit to be halted until he completes his business.)

Consequently, the experienced director will have his telephone calls screened and will clear up pending personal and business matters before his edit sessions. In addition, he will plan on some extended hours during offline edit sessions to take advantage of the "learning curve" that is inherent in the process. In other words, the

---

*Note: Offline editing generally refers to preliminary post-production sessions. Online editing is the final editing session in which the edited master tape is assembled from the original production footage. Editing techniques are discussed in detail in *Video Editing and Post-Production: A Professional Guide,* by Gary Anderson, Knowledge Industry Publications, Inc. Also see Appendix 11A at the end of this chapter for a brief discussion of offline and online editing.

ideal offline session is a long session with few interruptions. Fewer long sessions are more productive than many short sessions.

Organizational Strategies

A well-organized editing session allows the director to see a great deal of material in a short period of time. This can lead to creative opportunities for enhancing the program as he allows himself to concentrate on the program to the exclusion of other matters.

Use the Annotated Script

As discussed earlier in this chapter, I have found it most useful to approach an edit session with the minimum amount of necessary paperwork—simply the script with its notations of which takes to review. This allows me to stay focused on the program edit, unconcerned with any of the details of the production that are not needed. I will only refer to the linear camera notes if the selected takes don't seem to work or I suddenly recall a reason for reevaluating a selected take. In this fashion, I am able to move through the program edit quickly, knowing that after completing the rough cut I can still come back to any given point and recut segments.

Cue Time Management

In proceeding through the edit I make a point of using search times to allow me to review the program up to the point where I am editing. In other words, if I have sent a tape to a cuing point that I know will take two or three minutes of shuttle time, I will use that time to look at the last two or three minutes of the program. That way, I am constantly placing each edit into the overall context of the program, rather than dealing with it in isolation. If at any point in the program I find that an edit simply seems to not work, I will move on to the next called-for shot and look at it prior to returning to the problem edit. This often gives me the fresh perspective that I need in order to solve the problem.

Avoiding Predictability

One of the most engaging aspects of a good program's editing is that it will surprise viewers often enough to keep their attention. This is not accomplished by gratuitous tricks but by breaking predictable patterns. There are several techniques that allow a director to introduce the "surprise factor" into the edit process.

It is a good idea to periodically review the entire program, not just the last few minutes, up to the edit point, so that the timing and pacing within the program are clear. Although the timing of specific edits is sometimes dictated by the action from one shot to another, many other edits, such as the start of a scene or the end of a scene, can be cut at many different lengths. The review process allows me to judge whether to extend or compress these scenes to develop larger rhythmic patterns than those that arise from edit to edit.

Let's suppose I am editing certain sequences very tightly in an effort to keep the pace of a program active and thus develop a pattern of rapid edits. I might unintentionally create a staccato rhythm by having many quick edits in a row. When I review the program I will realize this and may adjust the timing of the edits to make the pattern less predictable.

"Split" Edit Options

The "Audio-Lead-Video-Follow" Edit

A common predictable pattern in dialogue scenes is to cut to each person when he or she speaks. This is often a matter of convenience; it is easier to perform one audio-plus-video edit than one audio edit followed by one video edit. However, this not only creates a predictable edit pattern, it is also not consistent with the way in which most viewers perceive the world around them. Most of our conversations involve some element of people talking over each other's lines and interrupting each other. A program that is all audio-plus-video edits is not consistent with those natural patterns.

To recapture the spontaneity of normal speech patterns (and to be able to surprise the viewer), I have found it very useful to create split edits in conversational program segments. In other words, if a character named Jack is speaking and the next person to speak will be Linda, I will allow Linda's words to begin before we see Linda speaking. Her audio edit will begin before her video edit, while Jack's face is still on the screen. This will often feel more natural than an audio-plus-video edit because, in fact, we often don't turn to see who is speaking until they begin to speak and we know in what direction to look. This is commonly referred to as an "audio-lead-video-follow" edit.

This type of split edit can also give a program a "live" feel rather than a staged feel. In live television, without a script, a director will always follow the speakers with their shot on camera after they begin to speak, rather than at the instant of their speaking. Human reaction time requires this. Viewers are subconsciously aware of this, and the director can use this fact in the edit process to inject that live feeling in a program.

"Reaction" Lead Edits

There is another form of split edit which also is patterned after the manner in which people relate in the real world. This type of edit also allows the director and editor to create a less predictable editing pattern.

Often a statement is made in a conversation that we know has a great impact on the listener. For example, Linda may tell Jack that she is considering firing him for a number of reasons and begin to enumerate them. As a viewer, I naturally want to see how Jack is reacting to this new information, even though Linda hasn't finished speaking. As a director I may choose to cut to Jack as soon as Linda's intent to fire him is made clear. Then I can remain with Jack on-screen until he speaks, which will undoubtedly be soon. This type of edit is called a "video-lead-audio-follow" edit.

Shooting for Edit Options

By utilizing edit techniques such as these I am able to break out of predictable edit patterns while staying within the accepted conventions of human behavior. The net effect is to enliven a program without making it seem staged or artificial. However, to have these edit options available, a director must have observed some fundamental rules while shooting.

Any split edit requires that the director provide the editor with the necessary "handles" on takes to have the room to make such edits. This simply means that the director should overlap the end of one take with the beginning of the next take, recording the off-camera speech that leads into the on-camera speech. He must also ask his performers to be in character while they listen prior to speaking and to remain in character after they finish speaking and begin listening. This is something that becomes second nature to seasoned professionals but may require some explanation with inexperienced actors or with nonprofessional talent.

Planning for Effects

In the course of developing a script, rehearsing and directing his talent, a director may decide to use specific visual effects in his program. The purpose of these effects should be to move the program along from one point to another in a fashion that is understandable to a viewer and has a different impact than a simple cut.

The most common effects that have been utilized are fades and dissolves. They commonly tell us that we are leaving the frame of real time and are moving ahead to another time and/or place. However, such effects have a language of their own. If done in a jarring fashion, the viewer may feel that the program's director or editor needs to hide something.

To ensure that these effects work, the director must plan his key actions within a scene to conclude prior to an effect or to flow naturally into an effect. This is often easy to see in the offline or rough cut stage by previewing a simple cut at the transition point. If the cut seems excessively abrupt, then the transition point should be adjusted until it seems the least abrupt in relation to the actions on both sides.

A common wisdom has it that dissolves can be used for transition from one action to another when a cut doesn't work. This is not always the case, since screen convention has regularly been that dissolves indicate specific intentions on the director's part. In light of this, my guideline has been to make most of a program's transitions work without any effects during the offline stage, and to be very clear on why I choose effects in the online stage.

PREPARING THE ONLINE EDIT

In the time I spent as an online editor for a post-production company, I learned that there are many ways in which a post-production house tries to help its clients become successful with their projects. Post-production management knows that much of their business results from satisfied clients who return to work with them again and again. It is in their own self-interest to assist their clients in ways that may mean fewer dollars in the short run, but that guarantee that all-important return business.

Getting Good Service

Management knows that a client who can rely on a company to help him get his program done on time and on budget will be most willing to continue to work with that company. For this reason, management is often willing to provide free and/or discounted services to producers and directors who do productions on a regular basis. For example, an editor can be made available, often at no charge, to discuss the best way to organize material for an edit, to look at an edit design and suggest effects or transitions that are available. In addition, there are many tasks that are required for an edit session, but do not have to be performed during the session at the full edit room rate.

Some of these tasks include the transfer of graphics to videotape for editing, pre-programming character generator information, transfers of material from one format to another (1-inch to Betacam, for example, or film to tape), and searching for sound effects and library music. These tasks can often be done at rates less than half that of the edit room's rate.

The Post-House as a Partner

In addition, when producers and directors encounter technical problems with their original footage, the post-production house will most likely be quite happy to help troubleshoot the problem prior to the edit session and to assist in finding an appropriate solution. In this way, it will increase the productivity of the upcoming ing edit session and will be seen as a valuable resource to be counted on for future projects.

Post-production houses will tend to be more helpful (and generous in discussing downtime) when working with directors and producers who develop a reputation for estimating edit time correctly. The client who consistently books near the correct amount of time for a project will be valued, because he helps the post-production house to run more efficiently and smoothly.

Editing Guidelines

To help in this relationship I try to follow some guidelines which have assisted me in creating good working relationships with post-production houses. First, I keep track of how many edits are in each program that I do. I also keep track of how long each program actually takes to edit during the online session. This allows me to assign a number-of-edits-per-hour figure to each project. I can then analyze each edit project to find out exactly why the number of edits per hour varied from my estimate.

This, in turn, lets me refine my estimate each time I supervise an edit session so that I can get closer to a correct figure each time, or at least pinpoint where I didn't allow enough time or allowed too much time. This makes me a better client for the post-production house and strengthens my relationship with it.

Some factors which come into play are the number of graphics and special effects in the program, how much audio "sweetening" is required, how many final client representatives will be present at the edit session and what their involvement is, and, quite simply, the difficulty of some of the edits. For example, very tight audio edits of conversations sometimes require multiple attempts before the right timing is established. With such tight edits, an offline edit system may not be adequate to establish final timing and some experimenting must be done online.

Communicating with the Editor

In preparing for an online edit session, the director should bear in mind that the editor who works on the program will be approaching the edit session from a very different knowledge base than the director. Whereas the director will have had long hours of living and breathing the program, to the editor, it is a completely new program about which he knows very little. Consequently, it is best that the director attempt to organize the elements for the edit session in a fashion that allows the editor to study them quickly and to understand the director's intent as swiftly as possible.

It will be helpful to create a detailed edit list with all edits' in and out times. In addition, the director may be able to bring the rough edit to the session and, providing the program is relatively short (under 15 minutes), record it onto the master as a guide, much like a scratch track is used for audio purposes. In the process of recording the rough cut to the master tape, the editor will see the entire program and cannot help but become familiar with it.

Editorial Assists

A copy of the final master script should be provided to the editor, particularly if a narration track is to be edited. If SMPTE time code is used as a reference for edits, the editor should have a copy of the edit list so that the director does not have to communicate the times verbally, which can lead to many mistakes.

The written time codes will allow the editor to enter the appropriate numbers on his first attempt. If an edit listing program is used, the director should make extensive use of text notation, so that the editor can see that a given edit is a close-up of a hand or a wide shot of a vista, in addition to a sequence of SMPTE numbers.

Language Conventions

A director who desires a smooth edit session should learn the conventions of any given edit system so that he is able to give an editor the appropriate information at the appropriate time for entry. (For example, some systems require that a dissolve be noted prior to entry of the in and out points of an edit rather than after. Knowing this, the director can save the editor many unnecessary key strokes.) Also, by using the correct terminology that the editor is used to, the director will reduce the amount of confusion that might otherwise occur if he were to use more popular expressions for effects and transitions.

By giving the editor the "big picture" at the start of the edit session, the director will be better able to enlist the editor's opinion and rely on his judgment. Otherwise, the editor will be very reticent to volunteer an opinion about a program that he is only discovering edit by edit as it comes together. It will also help the editor to move the edit session along if the director gives specific information regarding his priorities for the program—where he would like to spend time on details and where he feels that time need not be spent.

Procedural Guidelines

Specifically, the director might indicate whether he feels it necessary to preview and/or review edits or whether he will rely on the editor's confidence in viewin edits only as they occur before moving on to the next edit. [This is known as viewing in the E-to-E (electronics to electronics) mode and saves any re-cueing and reviewing that costs online edit time]. Since the editor is ultimately responsible for the technical quality of the product of the edit session, the director should involve the editor in this decision and defer to his judgment.

The director should also inform the editor as to the likelihood of a future reedit. The editor may wish to keep more extensive notes regarding effects and equalization levels if he knows that the program is likely to be reedited. These are on-the-spot adjustments that are not always noted by editors because of the time required to do so. However, in the event of a reedit, such notes can save valuable time.

Creating a Resource

Just as in selecting crew members, a director often develops personal favorites among editors at post-production houses. Over time, a good working relationship can be developed that benefits the director, because he is able to spend less time explaining his working preferences to the editor and more time focusing on the job at hand. Post-production houses are often willing to adjust schedules so that their clients can work with the editor of their choice, and most editors appreciate the opportunity to work with a client who has specifically requested them.

Since most edit facilities have sales representatives who regularly speak with the client, this provides another opportunity to cement the director-editor relationship by mentioning that you enjoy working with a particular editor for this or that reason. The editors who know that you have praised them to management are more likely to become personally involved in your project the next time you work with them.

As in so many other aspects of the production process, the successful director is one who recognizes the human factor involved and seeks to maximize the skills of the people with whom he works.

APPENDIX 11A: OFFLINE AND ONLINE EDITING

The following is a brief explanation of the technical aspects of offline and online video editing.

SMPTE (Society of Motion Picture and Television Engineers) time code that is recorded with sound and picture provides an absolute reference for a given video frame and makes an offline/online sequence possible. After material is recorded at the studio or in the field on "camera original" tapes, copies can be made which include the time code from the original in a window superimposed over the picture. Time code can also be placed on the audio track or the address track of the copy or on any combination of the three. This allows an editor to use copies of the original footage to make content and timing decisions to create an edited master from the copies (window dubs) of the original material.

Since the time code information from the original is included in this process, it lets the editor and director edit a program without risking damaging the original tapes. Later, after this offline edited master (the rough cut master) is approved, editor and director can use the original tapes to "conform" an edit. That is, they can create an edited master from the original tapes in which the edit points exactly correspond to those in the offline edit by referencing to the SMPTE time code.

For instance, a program may be shot on Betacam SP camera original videotapes. As soon as the production wraps, the original tapes are taken to a post-production facility where window dub copies of the original tapes are made. The original tapes include picture, audio track one, audio track two and a third audio track that has the SMPTE time code recorded as an audio signal. VHS window dubs are made, which include picture, audio track one and audio track two as mirror images of the original footage. In addition, the third audio track's SMPTE time code information is translated to a visual code that is superimposed over the picture (hence the term "window dub").

The VHS dubs are then used as source tapes in an inexpensive VHS-to-VHS edit system and director and editor create their offline master. After it is approved, and any revisions are included, the information in the window is used to create an edit decision list (EDL). This list serves as the guide for the online edit session.

For the online edit session, the post-production house retrieves the Betacam SP camera originals, loads them into the appropriate VTRs and edits on to a 1-inch master, following the EDL. Whereas, during the offline edit only edit timing and content decisions are important, during the online session, color and signal timing information is matched from scene to scene through time base correctors, processing amplifiers and color correctors. Audio levels and equalization is also matched and audio is "sweetened" for the best match throughout the program. Dissolves, wipes, special effects and graphics are also included as part of this process.

The result is a finished, polished program. (The process described above is very similar to the work-print edit phase, followed by a negative-cutting conformation phase that is traditional with film post-production.)

There are currently several alternative methods for offline editing, although the one described above is the most common. Three-quarter-inch to ¾-inch systems can be substituted for the VHS system described above. These systems are described as linear systems since one-for-one copies of originals are made and editing is done in a traditional (to video) linear fashion.

Some faster systems are gaining popularity. These systems utilize multiple copies of camera originals loaded in multiple source machines (Montage and Ediflex), or laser videodisc copies of the original material (CMX 6000 or Editdroid) to decrease the access time required. The basic priniciple remains the same. The offline editing process is basically an edit timing/content decision-making process in which editor and director do not concern themselves with technical considerations, since this consumes time unnecessarily. In the online process virtually all content and timing decisions have been made and editor and director can concentrate on the technical aspects of the edit.

12 The Review Process

THE VALUE OF REVIEWS

The alert director will value the review process in a larger framework than that of the specific project that it concludes. It can serve as a great help to him in developing his directorial capabilities as a professional in a challenging field.

In the review process of a program, in the aftermath of a production, he can look over his own work in light of his knowledge of the production's opportunities and limitations. He can then ask himself, "How'd I do this time out and what lessons can I take with me from this project?" Since he is the only person who fully knows what his initial vision of the finished program was, he is in the best position to judge the final result and gain in knowledge of his own working methods. He has the unique opportunity to use this knowledge to refine his craft.

Pre-Production Review

I have found it helpful to break the review of my own work down into several steps. Prior to each group of production days, I will take a look at how organized I am, going into the shooting phase of the program. There will always be some areas where I know exactly what I would like to achieve and some areas where I don't feel that I have a clear vision of what I would like to see happen. If I look more closely at the latter, I will find that some aspects are clouded because I haven't been given enough information, and some aren't clear because they involve uncontrollable or unpredictable production logistics. These I may not be able to do anything about.

However, there may be some aspects which are unclear because I haven't spent enough time researching or trying to visualize them. These represent areas in which I "haven't done my homework." These areas pose directorial problems in the making—areas to which I need to give additional attention as the project enters the production phase.

The "Dailies"

In the ideal production schedule, time is allowed to review footage at the end of each shooting day. It allows the director to see what is working on tape and what isn't. This is particularly important to a director who is coordinating many aspects of a production since he will not always have the appropriate "distance" from the action in front of the camera to perceive whether it is working for the viewer.

Stepping back at the end of the day to view the "dailies" allows the director to establish this distance and to place himself in the viewer's shoes for a time. With this perspective, he can see whether the pacing is too fast, whether the talent seems credible, whether the humor is working, and whether his staging is supporting the script's content.

I have found this exercise particularly important when I am shooting while traveling to distant locations or in noisy and busy environments. These situations involve distracting circumstances that dilute my concentration during the actual taping. By regularly viewing selected takes at the end of the session, I am able to compensate for any temporary loss of focus during the production day.

The Pre-Edit Review

Before beginning the actual editing of tape, I find it useful to look through all the footage marked as "director's selects." This gives me the chance to see the good points and bad points of the final takes. Also, I am able to spot problems in some takes and can look for alternates among the unselected takes before beginning the edit. This is a good practice in that it has a pacing on its own, a less active mode than editing, in which the problems that are discovered are not then holding up the edit process.

By looking at the footage prior to the edit session, I am much more relaxed and patient. This allows me to be more analytical about what worked and what didn't. I am more apt to learn from my mistakes when I discover them in a private, unpressured atmosphere, rather than an edit atmosphere. In the edit environment I am more apt to feel the pressure of time, to feel that I should immediately concentrate on making an alternative work. This does not allow me to look closely at the source of the problem and learn from it.

The Value of Isolation

Not having an editor or a client viewing with me greatly reduces the chances of personal embarrassment and allows me to be honestly self-critical. This also allows me to discover any major technical problems that might be due to crew errors. With no one else present, I can protect the crew members from having other persons see their mistakes. (This is a delicate area and one in which I think that the director owes his crew a professional courtesy. By finding alternative takes or ways around problems, he maintains the professional image of the entire production team.)

Just as important is the possibility that what appears on tape to be a mistake by a crew member may, in fact, have been the result of poor communication by the director. For example, a director might neglect to completely block out movement in a scene with the sound recordist present. When the talent makes a movement that brings him "off-mike" because the soundman is caught off guard, this actually represents a directing error, not an audio mistake.

Novice clients don't believe that professionals make such mistakes and are often unforgiving when confronted with them. (The reason the new client doesn't think that such mistakes are made is very simple—mistakes are always edited out of the finished programs that the public views.)

Since this is the new client's first look "behind the scenes" and his first exposure to technical errors by professionals (other than his possible viewing of "blooper, bleeps and blunder" reels), the director should remain aware of the impact of them on the client. It greatly affects the client's perception of the professionalism of the production team.

MAINTAINING APPROPRIATE CLIENT INVOLVEMENT IN POST-PRODUCTION

As was mentioned in the previous chapter, many clients are especially eager to be involved as soon as tape is finished shooting.

Should the director risk offending the client by seeming to shut him out of the edit process? Or should he invite him into edit sessions even though there is little that the client might be able to add at this point? Inexperienced clients have little to offer to an experienced director and editor. Clients experienced in post-production will often recognize when they can be helpful and when they will simply be in the way and usually act accordingly without creating this dilemma.

I have found that the best strategy is to let the client know, at the start of a production, when and where he will be helpful, and in what ways. At this time I can usually explain that during the rough edit stage I like to work without interruption to make the best use of his production dollars, and that I will gladly show him the first cut before anyone else sees it.

In fact, I will be asking for his help at that stage to catch any problems or errors and to solicit his general reaction. This has the effect of reinforcing my role as director and of reestablishing the fact that I was hired to make the creative judgment in the production and editing phases of the project. I have found that the best approach is to keep all but the most experienced clients out of edit sessions.

In commercial work the dynamic is quite different. The agency producers, creative directors and art directors control post-production and typically oversee initial edit sessions, and they have the experience to do so. Directors are often not involved in the process or are invited as a courtesy on the agency's part. Some directors are hired by agencies for their continuing involvement in the edit process, but this is the exception rather than the rule.

Another good reason to keep the individual pieces of the show "under wraps" until the client views the rough edit as a whole is that the director can create a more exciting viewing situation than would occur if the client had already seen much of the editing. He can maintain the element of pleasant surprise that should accompany the first review of a rough cut.

Further, there are many little flaws in any production which might leap off the screen when a portion of a take or a part of a scene is viewed in isolation. Those flaws become invisible when the program is viewed as a continuous piece. However, if the client sees the defect during or prior to the edit process, he will invariably be distracted from the flow of the program. He will continually "rediscover" the flaw every time he views it. That one imperfection will have affected the client's entire perception of the program.

CONDUCTING THE OFFLINE REVIEW

Keep it Exclusive

It is generally better to conduct tape review and logging in private, involve the client in the rough edit stage only as necessary and hold a formal viewing of the rough edit. Often it is necessary to make sure that the client feels comfortable with this approach. This can be done by reassuring him that no one else will see the edited piece until he has seen it and has had a chance to make his comments and suggestions.

Some clients need to feel a sense of personal ownership in a program that was produced with their money or that involved their department. By giving them an exclusive preview, they still feel that the program is "theirs." This would not be the case if they saw it for the first time with other members of their company who were not so involved. An exclusive viewing ensures the client's comfort by making him continue to feel a part of the creative team.

Accommodate the "Gatekeepers"

The ultimate acceptance or rejection of a program may lie more in the perception of the "gatekeepers" within a company than in the actual merits of the program itself. Gatekeepers are the people who have the most influence on others' concepts of what is good or bad about the world outside their own corporation, or who are perceived by their fellow workers as possessing insight or special knowledge in specialty areas such as video production.

Whether this perception is deserved or not, the director who wishes to create good client relations for the future is well advised to pay attention to these phenomena. The director who projects an attitude of professionalism and self-confidence can foster such attitudes in the gatekeepers within his client base and enhance the chances for client enthusiasm for and acceptance of his efforts.

Prepare the Client

Many clients have higher expectations for their programs than their budgets merit. Their frame of reference may be high-budget feature films or broadcast television programs with elaborate production values. Although they may have been prepared by the director at the outset of the production to set their expectations according to their limitations, the excitement that accompanies the production process often obscures the original promises. In light of this fact, it is often best to redefine expectations prior to reviewing an edited program with clients.

Restate the Program's Objectives

It is helpful to remind clients of the original objectives of the program. This is the first step in recreating for them the specific assumptions on which the production was designed at the outset of the project.

This can be especially important in a corporate marketing environment because marketing goals often change rapidly. I have sometimes created programs that were exactly what a client had requested at the outset of a project and then found that the client was less than ecstatic about the final program—not because it did not meet the agreed–upon objectives, but because it did not meet some new objectives that were defined long after the program was scripted.

Restating original objectives is also helpful in introducing a program to an audience that includes members of a company who have had differing degrees of involvement in the creation of the program. Bringing everyone "up to speed" can help the program meet with greater acceptance.

Let's say that a personnel department has undertaken a video project with specific objectives. When the offline review is scheduled, the department feels they must invite senior management to ensure the program's successful acceptance in the company. However, the senior management may bring an assumption of different objectives to their viewing. Often they assume that the program should meet objectives that they would have set if they had made the program.

It is essential, in such a context, to inform (or politely "remind") the audience as a whole of the program's objectives. Otherwise, individual viewers with different expectations may take exception to the program's style and tone. They may find it inappropriate to their own assumed objectives. Any secondary objectives should be noted as well.

Restate the Intended Audience

For the same reasons, the director should remind the viewers of the audience for whom the program was made. He should define the primary audience as well as any secondary audiences. Creating this perspective will let the viewers judge the program on its merits, rather than their own sets of prejudices.

For example, a program for hourly employees on an assembly line will meet with greater acceptance if the viewers know that casting, costuming and scripting were all geared to that particular audience.

Explain How the Tape Will Be Used

The director should also note if a program is to be preceded by a trainer's or a personnel officer's personal presentation or accompanied by literature. The viewers of the rough edit should understand why some elements are not covered by the tape itself—that they are to be in a print package or presented by a speaker. This will enable the initial viewing audience to "fill in the blanks" and see the program within its intended overall context. Otherwise, they will tend to watch in a distracted fashion as they perceive that something important has been omitted.

Tell Viewers That This Is Not the Final Edit

In his screening of the program, the director should look for those technical aspects which may need particular explanation. Perhaps a particular scene involves the compositing of two elements, as in an extended dissolve, or a scene will utilize an uncommon special effect. These aspects of the program should then be explained in sufficient detail prior to the client's viewing so that the client can imagine the finished effect without further intervention by the director.

In addition, an offline edit or rough cut does not include many of the refinements that will appear in the final program. It is important to point out this fact to members of the initial viewing audience. Otherwise, they may misjudge the program's overall quality. The director should tell them what to look for and what not to worry about.

Usually, this includes advising them to look at the overall content of the program. Often it is important to describe the specific elements that will be added in the next editing step.

How to "Set the Stage"

A typical speech to precede the "unveiling" of an offline program might run as follows:

> "The program we're about to see was done as an introduction to new-hires of the Compleat Insurance Corp. The program's intent is to give them a sense of the history of the company. As you know, it is not intended as an academic history but is meant to give the new employees a good sense of the roots of the company and leave them with an overall feeling that they are part of a proud tradition.
>
> In the edited version that you will be viewing, we are using copies of the original footage so as not to risk damaging those camera originals. This allows us to make and remake our content edits until we are satisfied with them. We plan to use graphics at certain points in the program, and those places are represented in this edit by a black screen.
>
> The music that you will be hearing is representative of our chosen selections, but the "mix," or ratio, between voice and music is still in the rough stage. Also, there are many points in the program that may seem like abrupt transitions. These transitions will be accomplished with dissolves—one picture overlapping the other—to create a softer transition in the final edit.
>
> We'd like you to view it once from start to finish to get a sense of the overall impact of the program. Then we can review it step by step a second time to address any concerns you might have. I'll be able to stop and start the program during our discussion of these points."

This approach allows me to let the program play after my initial introductory speech and gives me the opportunity to observe the audience's reactions. This is important since I am interested in the direct reactions of the initial viewing audience as a prediction of the response of the final intended audience.

Use Print Supports

I have also found it helpful to create print supplements for the offline review so that the audience can refer to graphic information that could not be included in the offline edit. Figure 12.1 is a sample from a marketing tape for a software company. The key to this approach is to create a generic source tape to be used in the offline

Figure 12.1: Sample Cue Sheet for Offline Review

Graphic #1

A black screen with the words:

Creative Software

Graphic #2

Fully animated air-brush color rendition of information flow which will become a rapid flow, then a "hot spot" symbolizing a potential problem appears, then alternate information routes appear and the "hot spot" subsides.

Graphic #3

A black screen with the words:

More Accuracy

to which will be added:

On Time

to which will be added:

All the Time

at which point, all three lines fade to black.

---

stage that has a succession of graphic indicators, the first being "GRAPHIC NUMBER ONE," the next, "GRAPHIC NUMBER TWO," and so on. In Figure 12.1, the edited offline tape has a frame that says "GRAPHIC NUMBER ONE" at the place where the title will appear, "GRAPHIC NUMBER TWO" where the animation will appear, and so on.

By editing the appropriate pre-built graphic "place-holder" into the offline edit and passing out a printed guide, I am able to explain what is missing without having to interrupt or "talk over" the tape.

FOLLOW UP THE OFFLINE REVIEW

These are the basic points that I have made many times over the years in attempts to give a program the fairest possible critique by its initial viewing audience. This is important because I feel that, with the proper introduction, the corporate audience can provide very good feedback regarding specific strong and weak points of a program.

The corporate audience usually has greater contact with the ultimate audience and can be more sensitive to certain issues and stylistic considerations than I might be. I may only be exposed to their particular corporate culture during the brief period of time in which their program is created. The audience may have several decades' worth of experience with the company and is attuned to the nuances and expectations of the final audience.

Eliciting Valuable Input

After the first viewing of the program a director should engage the client in a frank discussion of his initial impression of the program as a whole. He should ask whether the program meets the objective that client and director agreed upon at the outset of the production, whether it will be well received by its intended audience, whether the client himself is pleased by the result.

It is important to do this before a second viewing in order to get an honest and forthright first appraisal by the client of the program as a whole. On the second viewing, the focus will be on details, and often a client who has trouble being frank about his like or dislike of the whole program will tend to focus on details and obscure his true feelings.

Probe for True Reactions

The director should probe for the client's real overall impression, because that is what will tell him whether the program is effective or not. The details that the client cites may not be the ones that have actually led to the forming of his judgment. The director needs to lead the discussion to find the true source of the client's conceptions.

For example, a technically oriented client may have little or no experience with judging acting performance, but elements of weak characterizations may lead the client to feel that the program is ineffective. Instead of addressing the acting issue, with which he has little familiarity, the client may instead zero in on technical details of which he has a detailed knowledge.

It is important for the director to seek "hidden" objections so that he can attempt to address them in the immediate project. But it is also important for him to develop a better understanding of his client through those objections. This will enable him to build a better long-term working relationship with the client.

Avoid "Oversell"

The director is often tempted to convince the client that the program is absolutely terrific, to conduct a "sales job" on the client. This may help to win initial acceptance. But if there *are* problems with a program, and they are not addressed appropriately, they will return to haunt the director. Although he may do a good job of convincing the client for the moment, repeated viewings, particularly with fellow workers who might be brutally candid, will soon reestablish the program's problems in the client's mind.

By unearthing any concerns and reassuring the client at the initial viewing, the director can help the client deal with those same concerns should they be raised by co-workers at a later date. This will give the client renewed confidence in the client-director relationship and in the value of the program.

All such issues must be resolved before proceeding to the next step, the online edit session with its many associated costs. Once the program becomes a final edited master, client changes become very expensive to address.

For example, it is much less costly to schedule re-narration or to create a new graphic prior to the online session than to have to do the same after that session and, subsequently, have to create a new master tape and possibly new copies as well.

THE ONLINE EDIT SESSION

After the initial client review and comments, the director may have slight changes to make in a program and may need to provide the client with a second rough edit. In many instances the changes that are required can simply be discussed verbally. (It is often advisable to document this with a follow–up letter stating the content of the discussion and the plan of action.) Then, the director must prepare all elements for the final step: the online edit session.

In this phase all the final timing will be "carved in stone" as edits are performed on the program master tape. Effects will be added, dissolves performed and audio sweetened. In this step, the director is no longer concerned with the basic arrangement and pacing of scenes. Those content decisions were dealt with in the earlier phases of post-production. Instead, he should focus on the small details that add to a production's impact.

For example, he should have already made all major editing decisions (such as whether he is cutting out of a scene after one phrase, or another) and should be focusing instead on the rate of dissolves, how loud one character's lines should seem when they are off-camera, etc.

For this he needs extended concentration and good rapport with his editor. He also must make decisions quickly, since the session's hourly rate is usually quite high.

This is an arena in which an uncontrolled client can create unnecessary costs without knowing it. But since there is a regular pattern to such edit sessions, let me suggest some guidelines for involving a client without creating major problems.

Client Involvement in the Online Edit

If a client wishes to attend an online session, make clear to him ahead of time that there is an hourly clock ticking at the session and that it is important, for his program to come in on budget, to defer to the production demands at hand.

Often I will tell clients that they are not needed at the start of the edit session. This is a time that is spent setting up machines, establishing how certain video and audio problems will be handled, how scenes will be matched in color, etc. This work is essentially boring to the client who lacks technical background in video production. He will not see any real editing being done for the first hour or so and will get the impression that things aren't progressing on schedule.

In fact, every session has a slow start–up period that causes some anxiety on part of the paying customer. This is normal and expected. After the edit is underway and the editor becomes familiar with the program's organization, things begin to move relatively quickly. The faster pace makes up for earlier "lost time." The presence of the client at the beginning of the session intrudes on the development of rapport between director and editor, is distracting to both, and creates unnecessary tension.

Planning Client Presence at the Session

One good strategy is to ask a client to come to a session several hours after it is underway, so that there is a fair amount of program material already on the master tape when he arrives. This gives the client something to view and tends to give him a good impression of how his money is being spent. Also, this is when director and editor are ready to take a break and switch gears to the "client entertainment" mode.

If a session is an all-day edit session and the client is a major customer, the edit facility's client services representative may welcome an opportunity to take the client to lunch. If there is an established relationship between the director and the facility, this may be quite comfortable and appropriate for all parties. The client arrives at the edit session, views some finished work, and watches the edit session progress for a brief period. The facility's representative can then entertain the client for a good long lunch while the director and editor continue to work. After lunch, the relaxed client can view another stretch of the program.

I have found that most clients who are "handled" in this fashion get a clear picture that director and editor are engaged in serious business, are working in a very focused fashion, and that the result is quality work. Rarely does the client stay too long after the second viewing.

I favor such an approach with novice clients because the edit suite with its lights and equipment can be a very interesting place. There is also prestige attached to being able to say to one's fellow workers, "I'm off to the television studio for the day, I have to supervise an edit session." Without a strategy for controlling such clients, some potentially volatile situations can be created.

Obviously, each client should be handled differently according to his degree of experience with the process and ability to interact with the director and editor in an appropriate fashion. The director should control the situation with the client's best interest at heart, without ignoring the public relations aspect of involving the client in an intriguing environment.

LISTENING TO THE AUDIENCE

Being truly open to audience response may be one of the more difficult tasks for the director. But if he wishes to be successful with programs that are designed for specific audiences, he must remain open to the input that such audiences will offer.

After working very hard, the director may be very protective of "his" program and desire total unqualified acceptance by the audience. (Everyone likes to be complimented and directors are no exception.) But if the director allows his desire to please the immediate audience to get in the way of being a good listener, he may miss some valuable pointers that can help him to construct the next program for that audience.

For example, I once directed a marketing video for a high-tech telephone company. The program was to introduce a new product line and its accompanying marketing strategy to the company as a whole. The specific approach involved unveiling the program to company members and invited press representatives via a teleconference, an approach that was employed regularly by the company. At my request, I was able to attend one of the teleconference sites to observe audience reaction.

The company employees (primarily middle-management marketing types) chatted freely during the screening and "talked back" to their own representatives on the screen. They worked in an informal corporate atmosphere and showed a healthy irreverence for the formal presentation style that had been chosen. I realized from this experience that a more documentary style or a humorous presentation style could command greater attention within this context.

This aspect of the audience-specific nature of nonbroadcast television is one that makes it most intriguing for the director who is interested in the larger framework of communication strategies and their relative effectiveness.

Immediate Follow-up

The director should take advantage of every possibility for follow-up that is offered to him, and to look for other possibilities. One of the first opportunities will be the formal presentation of the video project by the company. If the director has the time to attend he can often learn specifics about the corporate culture within which his client works.

However, he should do so as a "fly on the wall"—an observer—rather than as the celebrity who was responsible for the production of the program. (The "Heisenberg principle" of the observer affecting the phenomenon is all too true in this instance.) He should strive to observe honest reactions rather than the response that a too-polite group might wish him to see.

Assisting the Client with Audience Acceptance

It will benefit the director in his client relations to tell his client how to position the program prior to showing it to others and to prepare the client for potential unenthusiastic responses. Often a program that is delivered to a round of initial praise later engenders negative reactions as colleagues who were not involved in its conception and production view the program without a proper introduction to its purpose and intended audience.

I once produced a program that demonstrated a blood-monitoring kit. It was a wonderfully compact and accurate device that helped diabetics and other persons with blood sugar problems to accurately measure their blood-sugar level in less than two minutes, which, in turn, allowed them to control their problem. The intent of the program was to ensure that purchasers of the system used it properly, since improper use would give them false results and false results would lead them to discontinue using the product. The program was a training tool first, and a sales aid in the longer view. It was not intended as a *direct* sales tool.

Upon viewing the tape, the client was quite pleased with the effectiveness of the presentation. However, her first company meeting after the completion of the tape was in another state and the majority of the attendees were from the company's sales staff. They objected to the program because they automatically assumed that any new video would be a powerful sales tool and expected the tape to have faster pacing and more of a sales message.

After hearing all their objections to the program my client was understandably quite upset. She called me with news of the negative acceptance. I explained to her that everyone brings his own expectations to the viewing of a tape and that she should review the objectives with any given audience. I then gave her an outline of how she might position the tape. Her next viewing met with a very positive response. Needless to say, I became a trusted source of advice after that and maintained a continuing relationship with the company.

Extended Follow-up

In the early stages of program planning, the director will probe for the intended short-term as well as long-term effects of a program. Often, he will want to know what the intended shelf-life of a program is. In order to find out how closely he met the longer-term goals and to see how a program stands the test of time, he should occasionally check in with his client to see how the program is received long after its initial production.

Many clients are pleasantly surprised by this display of long-term interest. It enhances the director's reputation and creates an even greater trust with his clients.

By making the effort to express my genuine interest in the continuing usage of a program I have created many lasting friendships with clients. I have received many "word-of-mouth" referrals from past clients that have allowed me to begin new client relationships on a firm footing of trust. This has led to some very interesting projects and has given me the opportunity to continue to work with appreciative clients—the perfect audience.

Afterword

In 1968 I began directing and editing programs using the first video systems available outside the traditional broadcast station environment. Starting with documentaries and progressing to training, marketing and commercial programs. I have found that directing video is one of the most challenging activities that a person who enjoys creative collaboration can pursue. The mix of artistic decision-making, technical problem-solving and people management is rare to find in any one career.

Crafting successful video programs is extremely satisfying in and of itself. But I have been able to gain a great deal of additional satisfaction from my work. I have made great professional friendships and assisted others in developing their own careers over the years. Building a network of mutually supportive professional relationships while watching others grow professionally and creatively has been as great a source of satisfaction as any of the plaudits and awards that I have received.

Every production offers unique lessons in human interaction and insights to the creative process. Taking the time to reflect on these aspects of past productions allows one to grow both professionally and personally. A director's education is an ongoing process. By integrating past experiences a director can control production logistics to better convey the meaning of each program he or she works on and, at the same time, create a distinctive personal style.

This book is a direct result of some seven years of seminars which I have conducted on the subject of directing. The response to those seminars has proved that there are some basic approaches which are not widely taught in traditional communications courses, but which people working daily in the production realm have found extremely valuable.

The preceding pages contain a combination of basic human psychology and practical business sense tempered by a technical awareness of the video production process. It is an attempt to convey those lessons and insights which I have gained over the past two decades. I hope that readers of this book have found them useful in their development of a directing style.

Thomas Kennedy
1989

APPENDIX A: SAMPLE PROGRAM PROPOSAL

A Proposal to HEAVY DUTY CLOTHING COMPANY (HDCC)

for a Retail Presentation Videotape on

HEAVY DUTY JEANS FOR COWBOYS

Basic Concept

This program will present the features of Heavy Duty Jeans (HDJs) for cowboys in a high-impact manner that will emphasize their down-to-earth design history. HDJs will be positioned as the choice of professional rodeo cowboys who make serious demands on their equipment—our "tough customers."

The buyer who views this program will relate to the no-nonsense product description; however, the presentation will utilize classic lighting and montage editing techniques to "punctuate" the presentation and maintain continuing viewer interest.

Three basic picture elements will be combined. Extreme closeups of HDJs will be shot under highly controlled lighting conditions to emphasize their specific features such as front loop placement, straight-leg styling, etc. These shots will appear in the same montage with wider shots of rodeo cowboys' hands and faces as they prepare their equipment for competition. The third element will be the fast intercutting of live-action of competitors *in* action as they rope or ride some other tough customers—professional rodeo livestock.

One voice-over narrator will be chosen to deliver a believable "real-person" narration in the style and speech pattern of a non-assuming rodeo cowboy. Live-action audio effects and music complete the soundtrack.

Rough Treatment

We open on solitary close-up shots of cowboys preparing for rodeo competition. The "look" is of directional, natural light, possibly early morning. First we see our BULLRIDER as he CHECKS HIS WRAP AND TESTS HIS RIGGING. We see DETAILS of his HEAVY DUTY JEANS, his face and hands as he checks his gear. A voice tells us that "THERE ARE SOME PRETTY TOUGH CUSTOMERS IN THE RODEO BUSINESS." (This is spoken as if the cowboy is thinking to himself or talking quietly to one person. The mood is PENSIVE and LACONIC.

SUDDENLY, we cut to very fast action of a BULLRIDER LEAVING THE CHUTE on a fast-twisting BRAHMA BULL. The FAST SLAM of the CHUTE GATE underscores this shot as we CUT IMMEDIATELY BACK to the quiet pre-rodeo atmosphere that might be under the stands or in a stable or corral. Our narrator tells us that "THERE ARE A LOT OF LITTLE THINGS THAT MAKE THE DIFFERENCE BETWEEN FINISHING IN THE MONEY AND MISSING THE CUT." The camera reveals more of the BULLRIDER AND HIS HEAVY DUTY JEANS as the voice goes on to say that Heavy Duty Clothing Company asked a lot of professional cowboys "WHAT THEY WANTED TO SEE IN A PAIR OF GOOD WORKING JEANS." We see the SPECIFIC PRODUCT DETAILS as they are mentioned.

We see a pair of hands PULL A STRAIGHT-STYLED LEG OVER A SNAKE-SKIN BOOT, hands buckle, then clear, AN OVER-SIZED PRIZE BELT BUCKLE to reveal FRONT LOOP SPACING, etc. We now cut to other competitors. A CALF-ROPER pulls his rope out of its CUSTOM ROPE-BOX and CHECKS ITS LAY as he re-coils it. A BRONC-RIDER SNUGS HIS GLOVE and checks its fit, then rests it on his Heavy Duty Jeans. These shots are again inter-cut with short bursts of RODEO COMPETITION—the QUICK THROW OF THE CALF-ROPER, THE SNAP-BACK ACTION OF THE BRONC-RIDER.

Our voice-over COUNTERPOINTS the fast action with its matter-of-fact narration of the story of Heavy Duty Jeans, telling us how HDCC LISTENED to its toughest customers to give them the FEATURES THEY WANTED.

We then "pay off" the sequence with a brief montage of the FULL ACTION of which we've only caught glimpses to this point. A complete 8-second BULL RIDE, 10-second SADDLE-BRONC RIDE and a complete 12-second CALF ROPING. This utilizes LIVE SYNC-SOUND and no voice-over to create an AUDIO BRIDGE to our CLOSING VOICE-OVER which calls out HEAVY DUTY JEANS' GENERAL FEATURES such as FABRIC, COLOR and FIT OPTIONS. We see samples of specific product options.

We conclude with a final live-action rodeo shot with HDCC's logo and a voice-over tag line such as:

"Heavy Duty Jeans—the choice of some pretty tough customers"

Budget

The cost for this program, exclusive of applicable tax (if necessary) is $28,180. This includes all pre-production, production and post-production, talent fees, permits and insurance and includes the delivery of one 1-inch videotape master. One draft script, one final script, an edited rough cut and soundtrack production are included as well.

A second budget of $20,550 is included at the client's request. This is for a program version which does not include the production of live-action rodeo footage. To sustain this program concept, black-and-white still photographs of rodeo competition might be substituted for the live action footage suggested in the first alternative. In either program approach, a full-crew, fashion-photo style video shoot is planned to capture the personalities of the professional rodeo cowboys. Estimates are based on sessions with 3 to 4 cowboys.

Additional program applications can be accommodated by increasing the shooting time at the rodeo, adding appropriate pre-production and post-production costs and covering applicable talent fees. A basic informational video without additional scripting could be produced for under $10,000. (Budget specifics will follow program guidelines as they are presented by the client.) A point-of-sale videotape would benefit from additional graphic treatment, as well as increased layering of images to sustain interest in a retail environment. A detailed, separate treatment and budget will be provided if this is an elected option. (See sample project budget on page 190.)

HEAVY DUTY CLOTHING COMPANY—PROJECT BUDGET

DATE: July 4, 1988

PROJECT NAME: HEAVY DUTY JEANS—Rodeo Cowboys—Reno & San Francisco

Pre-Production	Reno and San Francisco	San Francisco Only
R&D	$ 1,100.00	$ 700.00
Script/Outline	2,000.00	2,000.00
Secretarial Services		
Printing		
Graphics		
Photos/Film/Slides	500.00	500.00
Tape—Audio/Video	500.00	320.00
Sets		
Props	300.00	300.00
Costumes	150.00	150.00
Music/SFX	1,200.00	1,200.00
Subtotal:	$ 5,650.00	$ 5,350.00
Production		
Studio		
On Location	$ 150.00	$ 150.00
Outside Facilities	250.00	250.00
Misc. Location Insurance	400.00	100.00
Subtotal:	$ 800.00	$ 500.00
Equipment		
Studio Equip.		
Portable Equip. Dolly	$ 210.00	$ 210.00
Rental Equip. Grip & Lights	650.00	650.00
Overhead/Maintenance		
Misc. Expendables/Filter Rental	185.00	185.00
Subtotal:	$ 1,045.00	$ 1,045.00
Post-Production		
a. Tape Review	$ 400.00	$ 400.00
b. Editing	1,140.00	1,140.00
c. Tape (audio/video)	370.00	370.00
d. Video Dubs	100.00	100.00
e. Duplication		
f. Packaging		
g. Titling, Graphics Cam. & Mixdown	620.00	620.00
h. Misc.		
Subtotal:	$ 2,630.00	$ 2,630.00

HEAVY DUTY CLOTHING COMPANY—PROJECT BUDGET (Cont.)

Personnel					Reno and San Francisco	San Francisco Only
a. Producer/Director Pre-Pro, Shoot					$ 6,800.00	$ 4,000.00
b. Art Director					2,000.00	2,000.00
c. Production Crew					5,150.00	2,775.00
1. Sound Recordist	(3)	375.00	@	day (0)		
2. Technician	(3)	275.00	@	day (1)		
3. Gaffer	(2)	375.00	@	day (2)		
4. Grip	(1)	275.00	@	day (1)		
5. Production Assistant	(2)	125.00	@	day (2)		
6. Other P. Co-ord.	(6)	175.00	@	day (2)		
"Groomer"/Stylist	(2.5)	350.00	@	day (2.5)		
d. Talent					560.00	975.00
e. Editor					1,700.00	1,700.00
f. Misc.					130.00	130.00
Subtotal					$16,340.00	$10,675.00

Expenses	Reno and San Francisco	San Francisco Only
a. Transportation	$ 530.00	$ 200.00
b. Lodging	675.00	0.00
c. Meals	510.00	150.00
d. Misc.		
Subtotal:	$ 1,715.00	$ 350.00
Total:	$28,180.00	$20,550.00

APPENDIX B: SAMPLE BID PROPOSAL

TEK Productions
Production Avenue
Production City, CA 94960
(415) 555-5555

CLIENT: Heavy Duty Clothing Company
One Apparel Building
San Francisco, CA 94120

TELEPHONE: (415) 555-5555

PROGRAM(S):

1.   Tearing Resistance Test Procedure
2.   Breaking Load (Grab Method) Test Procedure

BID DATE: July 31, 1987
BY: Tom Kennedy

To produce two (2) videotape programs, each approximately 5:00 to 7:00 minutes in length. Visual elements to be recorded at HDCC's factory, utilizing HDCC personnel as on-camera talent. Professional voice-over narration, music and graphics to be combined with this element in post-production.

Crew to include: Director/camera, lighting/grip, VTR operator/technical director, production assistant. Total crew: 4 persons.

Pre-Production & Wrap Costs	$ 1,980.00
Shooting Crew Labor	810.00
Location & Travel Expenses	110.00
Props, Wardrobe, Animals	75.00
Studio & Set Construction Costs	— 0 —
(LS & Co Equipment—Cam/VTR & Off-line FAX	1,375.00)
Addn'l. Equipment Costs (Lenses, addn'l. lights)	425.00
Tape Stock & Window Dubs	350.00
Miscellaneous	80.00
Director/Creative Fees	750.00
Insurance	230.00
Talent Fees & Expenses	850.00
Editorial & Finishing	7,430.00
Scriptwriting	1,500.00
TOTAL	$15,965.00
(TOTAL with HDCC equipment provided	$14,590.00)

All costs to be directly billed to HDCC—direct sales tax included in above costs, no additional tax included.

APPENDIX C: DIRECTORY OF GUILDS AND UNIONS

Actors and Artists

American Federation of Television and
    Radio Artists (AFTRA)
1350 Avenue of the Americas
New York, NY 10019
(212) 265-7700

Screen Actors Guild (SAG)
1515 Broadway
New York, NY 10036
(212) 944-1030

Screen Extras Guild (SEG)
3629 Cahuenga Blvd.
Los Angeles, CA 90068
(213) 851-4301

Technicians

International Alliance of Theatrical
    Stage Employees (IATSE)
14724 Ventura Blvd.
Sherman Oaks, CA 91403
(818) 905-8999

International Television Association (ITVA)
3 Dallas Communications Complex
6311 N. O'Connor Rd., Suite 110
Irving, TX 75039
(214) 869-1112

National Association of Broadcast Employees
    and Technicians (NABET)
333 No. Glenoaks Blvd., Suite 640
Burbank, CA 91502
(818) 846-0490

North American Television Institute
    at Video Expo
Knowledge Industry Publications, Inc.
701 Westchester Ave.
White Plains, NY 10604
(914) 328-9157

The Sony Institute of Applied Video
    Technology
2021 North Western Ave.
P.O. Box 29906
Los Angeles, CA 90029
(213) 462-1987

Directors

Directors Guild of America (DGA)
110 W. 57th St.
New York, NY 10019
(212) 581-0370

Writers

Writers Guild of America, West (WGA)
8955 Beverly Blvd.
Los Angeles, CA 90048
(213) 550-1000

Transportation

Studio Transportation Drivers Local 399,
    International Brotherhood of Teamsters
4747 Vineland Ave., Suite E
North Hollywood, CA 91602
(818) 985-7374

Sound

Broadcast TV Recording Engineers
    International Brotherhood of
    Electrical Workers (IBEW)
3518 Cahuenga Blvd. W., Suite 307
Los Angeles, CA 90068
(213) 851-5515

International Sound Technicians,
    Cinetechnicians and TV Engineers
    Local 695, IATSE
11331 Ventura Blvd., Suite 201
Studio City, CA 91604
(818) 985-9204

Sound Construction Installation &
    Maintenance Technicians Local 40
    (IBEW)
5643 Vineland Ave.
North Hollywood, CA 91601
(818) 877-1171

Musicians and Composers

American Guild of Authors & Composers/
    The Songwriters Guild
6430 Sunset Blvd., Suite 1113
Los Angeles, CA 90028
(213) 462-1108

American Guild of Musical Artists (AGMA)
12650 Riverside Dr., Suite 205
North Hollywood, CA 91607
(818) 877-0683

Musicians Union Local 47 American
    Federation of Musicians/AFL-CIO
817 No. Vine St.
Los Angeles, CA 90038
(213) 462-2161

Talent Agencies

Association of Talent Agents
9255 Sunset Blvd., Suite 930
Los Angeles, CA 90069
(213) 274-0628

Wardrobe and Makeup

Association of Film Craftsmen Local 531
    (NABET)
1800 No. Argyle St., Suite 501
Los Angeles, CA 90028

Costume Designers Guild Local 892
    (IATSE)
14724 Ventura Blvd., Penthouse
Sherman Oaks, CA 91403
(818) 905-1557

Motion Picture Costumers Local 705
    (IATSE)
1427 No. La Brea Ave.
Los Angeles, CA 90028
(213) 851-0220

The location, addresses and phone numbers of regional and East Coast offices of all of the above guilds and unions are available by inquiring to the offices listed here.

APPENDIX D: RESOURCES

Recommended Books

Anderson, Gary H. *Video Editing and Post-Production: A Professional Guide,* 2nd edition. White Plains, NY: Knowledge Industry Publications Inc., 1988.

Bare, Richard L. *The Film Director: A Practical Guide to Motion Pictures and Television Techniques.* New York: Macmillan Publishing Co., 1971.

Field, Syd. *Screenplay: The Foundations of Screenwriting.* New York: Dell Publishing Co., 1984.

Glenn, Stanley. *A Director Prepares.* Dickenson Publishing Co., Inc., 1973.

Gradus, Ben. *Directing: The Television Commercial.* Stoneham, MA: Focal Press, 1981.

Johnson, Albert and Johnson, Bertha. *Directing Methods.* San Diego, CA: A. S. Barnes and Co., 1970.

Le Tourneau, Tom. *Lighting Techniques for Video Production: The Art of Casting Shadows.* White Plains, NY: Knowledge Industry Publications, Inc., 1987.

McMullan, Frank. *The Directorial Image.* Hamden, CT: The Shoe String Press, Inc., 1962.

Medoff, Norman J. and Tanquary, Tom. *Portable Video: ENG and EFP.* White Plains, NY: Knowledge Industry Publications, Inc., 1986.

Millerson, Gerald. *Basic TV Staging,* 2nd ed. Stoneham, MA: Focal Press, 1982.

Moore, Sonia. *The Stanislavski System,* 2nd rev. ed. New York: Penguin Books, 1984.

Morrison, Hugh. *Directing in the Theater.* New York: Theatre Arts Books, 1984.

O'Brien, Mary Ellen. *Film Acting: The Techniques and History of Acting for the Camera.* New York: Arco Publishing Co., 1982.

Silver, Alain and Ward, Elizabeth. *The Film Director's Team: A Practical Guide to Organizing and Managing a Film Production.* New York: Arco Publishing Co., 1983.

Van Nostran, William. *The Nonbroadcast Television Writer's Handbook.* White Plains, NY: Knowledge Industry Publications, Inc., 1983.

Zettl, Herbert. *Television Production Workbook,* 4th ed. Belmont, CA: Wadsworth Publishing Co., 1985.

INDEX

ABOUT THE AUTHOR

Thomas Edward Kennedy was born in Portland, OR in 1947. He attended the University of Santa Clara Honors College where he received his B.A. in philosophy in 1969. He founded the first two local origination channels in Portland using portable videotape as a production tool and initiated the use of video for the Environmental Studies office of Skidmore, Owings and Merrill Colleges. Returning to the San Francisco Bay Area, he worked progressively as editor for a broadcast station, senior editor for a production facility, senior director and producer/director.

Mr. Kennedy's involvement with corporate video projects paralleled the rapid growth of videotape as a marketing tool for companies in California's "Silicon Valley." He has also directed commercials and national PBS series. His awards include international gold and silver reels for marketing and public affairs programs as well as citations from NASA for coverage of the Jupiter and Saturn encounters. Mr. Kennedy has conducted seminars on directing throughout the country for the North American Television Institute. He is now with northern California's largest production facility, One Pass Film and Video, in San Francisco, where he specializes in special effects programs and commercials.